A LETTER FROM PETER MUNK

Since we started the Munk Debates, my wife, Melanie, and I have been deeply gratified at how quickly they have captured the public's imagination. From the time of our first event in May 2008, we have hosted what I believe are some of the most exciting public policy debates in Canada and internationally. Global in focus, the Munk Debates have tackled a range of issues, such as humanitarian intervention, the effectiveness of foreign aid, the threat of global warming, religion's impact on geopolitics, the rise of China, and the decline of Europe. These compelling topics have served as intellectual and ethical grist for some of the world's most important thinkers and doers, from Henry Kissinger to Tony Blair, Christopher Hitchens to Paul Krugman, Peter Mandelson to Fareed Zakaria.

The issues raised at the Munk Debates have not only fostered public awareness, but they have also helped many of us become more involved and, therefore, less intimidated by the concept of globalization. It is so easy to be inward-looking. It is so easy to be xenophobic. It is so easy to be nationalistic. It is hard to go into the unknown. Globalization, for many people, is an abstract concept at best. The purpose of this debate series is to help people feel more familiar with our fast-changing world and more comfortable participating in the universal dialogue about the issues and events that will shape our collective future.

I don't need to tell you that that there are many, many burning issues. Global warming, the plight of extreme poverty, genocide, or our shaky financial order: these are just a few of the critical issues that matter to people. And it seems to me, and to my foundation board members, that the quality of the public dialogue on these critical issues diminishes in direct proportion to the salience and number of these issues clamouring for our attention. By trying to highlight the most important issues at crucial moments in the global conversation, these debates not only profile the ideas and opinions of some of the world's brightest thinkers, but they also crystallize public passion and knowledge, helping to tackle some of the challenges confronting humankind.

I have learned in life — and I'm sure many of you will share this view — that challenges bring out the best in us. I hope you'll agree that the participants in these debates challenge not only each other but also each of us to think clearly and logically about important problems facing our world.

Peter Munk
Founder, Aurea Foundation
Toronto, Ontario

DO HUMANKIND'S BEST DAYS LIE AHEAD?

PINKER AND RIDLEY VS. DE BOTTON AND GLADWELL

THE MUNK DEBATES

Edited by Rudyard Griffiths

ANANSI

This edition published in 2016 by
House of Anansi Press Inc.
www.houseofanansi.com

House of Anansi Press is committed to protecting our natural environment.
As part of our efforts, the interior of this book is printed on paper that contains
100% post-consumer recycled fibres, is acid-free, and is processed chlorine-free.

20 19 18 17 16   1 2 3 4 5

Library and Archives Canada Cataloguing in Publication

Do humankind's best days lie ahead? : the Munk debates/
Steven Pinker, Matt Ridley, Alain de Botton, Malcolm Gladwell;
edited by Rudyard Griffiths.

(The Munk debates)
Issued in print and electronic formats.
ISBN: 978-1-4870-0168-1 (paperback). ISBN: 978-1-4870-0169-8 (html).

1. Progress. 2. Civilization. 3. Forecasting. 4. Social prediction.
I. De Botton, Alain, panelist II. Pinker, Steven, 1954–, panelist
III. Ridley, Matt, panelist IV. Griffiths, Rudyard, editor V. Gladwell,
Malcolm, 1963–, panelist VI. Series: Munk debates

HM891.D64 2016   303.44   C2016-901305-7
                    C2016-901306-5

Library of Congress Control Number: 2016933753

Cover and text design: Alysia Shewchuk
Transcription: Transcript Divas

*We acknowledge for their financial support of our publishing program
the Canada Council for the Arts, the Ontario Arts Council, and the Government of
Canada through the Canada Book Fund.*

Printed and bound in Canada

CONTENTS

INTRODUCTION BY RUDYARD GRIFFITHS

Do Humankind's Best Days Lie Ahead? was a departure for our series. For going on almost a decade, the semi-annual Munk Debates have focused primarily on hot-button geopolitical, social, economic, or technological issues. So why all of a sudden did we decide that our autumn 2015 debate should have a strong philosophical bent? We believe that many of the larger debates we are currently having as a society are wrapped up in the timeless human quandary of whether or not we are progressing as a species.

Many commentators look at current turmoil in world events and see a slow and steady decline of the rules-based order and economic progress that defined the half-century following World War II. Others believe that the diffusion of the power of nations to non-state actors such as NGOs (non-governmental organizations) and global alliances and institutions—all supercharged by instantaneous communication and globe-spanning social networks—as heralding a new era of peace and prosperity. Proponents of a "second" technological revolution envision a future with soaring living standards, new realms of individual and collective collaboration and expression, and a cleaner, greener planet. Their detractors rail against how new technologies fuel social and economic inequality, empowering states to violate the privacy of their citizens, and distracting humanity from the hard work and sacrifice necessary to maintain a sustainable life on planet earth.

In short, for half of us, whatever the issue or challenge, the glass is half empty. For the other half of humankind, the glass is half full. Privately, we are either optimists or pessimists about the state and future of our own lives, our societies, and the larger human condition.

The Munk Debate on Progress was a conscious effort to push the more than 3,000 people who filled Toronto's Roy Thomson Hall to chose a side, once and for all, in this timeless debate of the modern era. To paraphrase the debate's resolution: Do humankind's best days lie ahead, or not?

Before the event began, the audience was asked to answer the question. At the end of the hour-and-a-half contest, the audience voted again. The fast-paced and highly entertaining debate in the subsequent pages with all of its rhetorical quips, attacks, and parries ended up changing more than a few minds. To find out if you think humanity's collective "cup" is indeed half empty or half full, read on. I guarantee this mind-expanding debate will hold you to its final word.

Rudyard Griffiths
Chair, The Munk Debates
Toronto, November 2016

Do Humankind's Best Days Lie Ahead?

Pro: Steven Pinker and Matt Ridley
Con: Alain de Botton and Malcolm Gladwell

November 6, 2015
Toronto, Ontario

DO HUMANKIND'S BEST DAYS LIE AHEAD?

RUDYARD GRIFFITHS: Good evening, everybody. My name is Rudyard Griffiths and I'm the chair of the Munk Debates. It's my privilege to have the opportunity to moderate tonight's contest.

I want to start this evening by welcoming the television audience across North America tuning into this debate, everywhere from CPAC, Canada's public affairs channel, to C-SPAN across the continental United States. A warm hello, also, to our online audience watching this debate right now on www.munkdebates.com. It's great to have you as virtual participants in tonight's proceedings. And, finally, hello to you, the more than 3,000 people who have sold out Roy Thomson Hall for yet another Munk Debate just weeks after our much acclaimed Canadian federal election debate. It's terrific to have all of you here tonight.

Tonight's debate is a bit of a departure for us. We're not going to be talking about a specific geopolitical issue or cultural debate. Instead, we're going to think bigger. We're going to reflect on the nature of our society, its most deeply held beliefs, all in the context of the question we're posing tonight: Is humankind progressing? Do our best days lie ahead?

To reflect on this big question, a debate that has raged in our society and our civilization for more than two centuries, we've brought four people here to this stage in Toronto — people we think are some of the sharpest minds and brightest thinkers in their respective fields.

But before we get to that, I want to mention that none of these debates would be possible without the generosity, support, and vision of our hosts tonight. Please join me in an appreciation of Peter and Melanie Munk and the Aurea Foundation. Thank you. Bravo.

Let's get our debaters out here on stage and our debate underway. Our resolution is, "Be it resolved: humankind's best days lie ahead." Please welcome, speaking for the "pro" team, Montreal native, pioneering cognitive scientist, and internationally renowned writer and scholar Steven Pinker.

Steven's teammate is a member of the British House of Lords. He's a storied journalist, a contributor to the *Times* of London, and the author of a string of big, internationally bestselling books on the intersections of evolution, ideology, history, and progress. We know him as Matt Ridley. Matt, come on out. Great to have you here.

Well, one great team of big-thinking debaters

deserves another. Please welcome the celebrated U.K.-based author, broadcaster, and thinker, one of the leading public philosophers of his generation, Alain de Botton.

Alain's debating partner is someone we love to read regularly in the *New Yorker*, where he's a staff writer. We've also read a few of his books. I hear there are over ten million in print. Ladies and gentlemen, Canada's Malcolm Gladwell.

Let's quickly run through our pre-debate checklist. First: our lovely countdown clock. This is going to keep our debaters on their toes and our debate on time. To those of you who are new to the Munk Debates, when you see this clock get to zero, please join me in a round of applause for our debaters, which will let them know that their allotted time has been used up.

Next, I want to review the poll results from the start of this evening. All 3,000 of you coming into this auditorium tonight were asked to vote on the resolution, "Be it resolved: humankind's best days lie ahead." The results are interesting: 71 percent of you agree, 29 percent disagree. The cup is definitely half full for this group.

But, as we know, these debates change, they're fluid. So we asked you: Depending on what you hear tonight, are you willing to change your vote over the next hour and a half? Ninety-one percent of you—yes, that high—would change their vote. Only 10 percent of you were committed optimists. So, we have a real debate on our hands.

I'm now going to call on our first opening statement of the evening, which will go to the "pro" team, as is custom. Steven Pinker, your eight minutes begin now.

STEVEN PINKER: Fellow Canadians, citizens of the world, I plan to convince you that the best days of humankind lie ahead. Yes, I said convince.

Declinists speak of a faith or belief in progress, but there's nothing faith-based about it. Our understanding of the human condition must not be grounded in myths of a fall from Eden or a rise to Utopia, nor on genes for a sunny or morose temperament, or on which side of the bed you got out of this morning.

And it must not come from the headlines. Journalists report plane crashes, not planes that take off. As long as bad things haven't vanished from the earth altogether, there will always be enough of them to fill the news. And people will believe, as they have for centuries, that the world is falling apart.

The only way to understand the fate of the world is with facts and numbers, to plot the incidence of good and bad things over time—not just for charmed places like Canada but for the world as a whole—to see which way the lines are going, and identify the forces that are pushing them around. Allow me to do this for ten of the good things in life.

First, life itself. A century and a half ago, the human lifespan was thirty years. Today, it is seventy, and it shows no signs of levelling off.

Second, health. Look up smallpox and cattle plague in Wikipedia. The definitions are in the past tense—"smallpox *was* a disease"—indicating that two of the greatest sources of misery in human existence have been eradicated forever. The same will soon be true for polio and guinea worm, and we are currently decimating hook-worm, malaria, filariasis, measles, rubella, and yaws.

Third, prosperity. Two centuries ago, 85 percent of the world's population lived in extreme poverty. Today, that's down to 10 percent. And according to the United Nations (UN), by 2030 it could be zero. On every continent people are working fewer hours and can afford more food, clothes, lighting, entertainment, travel, phone calls, data...and beer.

Fourth, peace. The most destructive human activity, war between powerful nations, is obsolescent. Developed countries have not fought a war for seventy years; great powers for sixty years. Civil wars continue to exist, but they are less destructive than interstate wars and there are fewer of them. This pin on my lapel is a souvenir from a trip I took earlier this week to Colombia, which is in the process of ending the last war in the Western Hemisphere.

Globally, the annual death rate from wars has been in bumpy decline, from 300 per 100,000 during World War II, to 22 in the 1950s, 9 in the seventies, 5 in the eighties, 1.5 in the nineties and 0.2 in the aughts. Even the horrific civil war in Syria has only budged the numbers back up to where they were in 2000.

Fifth, safety. Global rates of violent crime are falling, in many places precipitously. The world's leading criminologists have calculated that within thirty years we will have cut the global homicide rate in half.

Sixth, freedom. Despite backsliding in this or that country, the global democracy index is at an all-time high.

More than 60 percent of the world's population now lives in open societies, the highest percentage ever.

Seven, knowledge. In 1820, 17 percent of people had a basic education. Today, 82 percent do, and the percentage is rapidly heading to a hundred.

Eight, human rights. Ongoing global campaigns have targeted child labour, capital punishment, human trafficking, violence against women, female genital mutilation, and the criminalization of homosexuality. Each has made measureable inroads. And, if history is a guide, these barbaric customs will go the way of human sacrifice, cannibalism, infanticide, chattel slavery, heretic burning, torture executions, public hangings, debt bondage, duelling, harems, eunuchs, freak shows, foot binding, laughing at the insane — and the designated hockey goon.

Nine, gender equity. Global data show that woman are getting better educated, marrying later, earning more, and are in more positions of power and influence.

Finally, intelligence. In every country, IQ has been rising by three points a decade.

So what is the declinists' response to all of this depressing good news? It is: "Just you wait. Any day now a catastrophe will halt this progress or push it into reverse." With the possible exception of war, none of these indicators is subject to chaotic bubbles and crashes like the stock market. Each is gradual and cumulative. And, collectively, they build on one another. A richer world can better afford to clean up the environment, police its gangs, and teach and heal its citizens. A better-educated and more female-empowered world will indulge fewer autocrats and start fewer stupid wars. The technological advances that have propelled this progress will only accelerate. Moore's law is continuing, and genomics, neuroscience, artificial intelligence, material science, and evidence-based policy are skyrocketing.

What about the science fiction dystopias? Most of them, like rampaging cyborgs and engulfment by nano-bots, are entirely fanciful and will go the way of the Y2K bug and other silly techno-panics.

Two other threats are serious but solvable. Despite prophecies of thermonuclear World War III and Hollywood-style nuclear terrorism, remember that no nuclear weapon has been used since Nagasaki. The Cold War ended. Sixteen states have given up nuclear weapons programs, including, this year, Iran. The number of nuclear weapons has been reduced by more than 80 percent. And a 2010 global agreement locked down loose nukes and fissile material. More important, the world may only have to extend its seventy-year streak another few decades: a road map for the phased elimination of all nuclear weapons has been endorsed in principle by major world leaders, including those of Russia and the United States.

The other is climate change. This may be humanity's toughest problem, but economists agree it is a fixable one. A global carbon tax would incentivize billions of people to conserve, innovate, and switch to low-carbon energy sources, while accelerated research and development (R&D) in renewable energy, fourth-generation nuclear power, and carbon capture would lower their costs. Will the world suicidally ignore these solutions? Well, here are three *Time* magazine headlines from just last month: "China Shows It's Getting Serious about Climate Change"; "Walmart, McDonald's, and 79 Others Commit to Fight Global Warming"; and "Americans' Denial of Climate Change Hits Record Low."

A better world, to be sure, is not a perfect world. As a conspicuous defender of the idea of human nature, I believe that out of the crooked timber of humanity no truly straight thing can be made. And, to misquote the great Canadian Joni Mitchell, "We are not stardust, we are not golden, and there's no way we're getting back to the garden." In the glorious future I am envisioning there will be disease and poverty; there will be terrorism and oppression, and war and violent crime. But there will be much, much less of these scourges, which means that billions of people will be better off than they are today. And that, I remind you, is the resolution of this evening's debate. Thank you.

RUDYARD GRIFFITHS: Two seconds to spare. Steven, that was an impressive start to the debate. Alain, you're up next. Your opening statement, please.

ALAIN DE BOTTON: Thank you so much. If we're going to be optimists, like our learned friends on the opposing team, we're really going to look at distilling the themes that Steven and Matt are going to talk about down to four things. Let's focus on the four areas where optimists think we're going to make the largest gains.

First, they are believers in the victory of knowledge over ignorance. Ignorance, a big scourge of our times, will be resolved through the light of reason. That's the great hope of the optimists. They also think that the desperate ills of poverty that have accompanied us for so long will be wiped out through the growing economies of the world.

The third point, war. War will be eliminated by the rule of the law and the increasing monopolization of power by states that follow international regulations. And, lastly, they believe disease will be wiped out through that wonderful tool, medicine.

And with those four things—ignorance, poverty, war, and disease—under control, we will rise onto a sunlit highland that our optimist friends want to tell us is on the way.

I've got one major objection, and it's a slightly auto-biographical one. I'm Swiss, and I've spent quite a bit of time in the country. The thing about Switzerland is that it has solved all these problems. It's got a fantastic education system. The average salary is $50,000 a year. The country has been at peace since the Treaty of Westphalia in 1648. And the hospitals are superlative. Yet, it's not paradise. Indeed, there are legions of problems. Yes, I would describe them as first-world problems, but they are no joke. They are genuine issues.

Switzerland is where Swaziland, Botswana, and Liberia may be in five hundred years' time. But the bad news, ladies and gentleman, is that Switzerland is not perfect. And it's for that reason that we have to discount the optimistic tenor of the motion tonight.

Why are Switzerland and countries like it not perfect?

Well, first because idiocy is not removed by reason. The great promise of the Enlightenment was that if you tell people what the right thing to do is they will do it, that evil is the result of ignorance. It's not. Idiocy is more stubborn than that. Poverty is not eradicated by raising the gross domestic product (GDP). There are millionaires and billionaires who feel they don't have enough—and that is the true definition of poverty, the sense you don't have enough. And, unfortunately, that rises and is present at any income level. War is not the last word on meanness and violence and cruelty. These things continue in societies even though people are not bludgeoning each other to death.

And, finally, even though there's no smallpox or guinea worm in Switzerland, people still die, despite the wonderful advances of medicine. Death has not been eradicated. And as far as I know, though my learned friends have another view, there is no cure for it on the horizon. These are the problems that we face.

Now, someone may say that with machines, technology, the Internet, and the iPhone we will, perhaps, gather together and produce a creature who is perfectly wise, perfectly kind, and immortal. Maybe we will, but this person is not a human being. *Homo sapiens* are a different species. We will never be able to evolve out of these roadblocks that I've been describing to you. I believe that at the top of our spinal cords we have what I like to call a faulty walnut. A walnut mind that has very destructive impulses, and is immune to certain kinds of education, and that resists any attempt to help it in many situations.

I believe, ladies and gentlemen, that we must move

toward a different sort of philosophy that will serve us much better. And that philosophy I call pessimistic realism. It's a counter to the boosterish attitude you find in modern science and in modern business, which for different sets of reasons is permanently trying to get us to feel more cheerful about things.

This kind of boosterishness is dangerous and cruel. Think of its application in relationships. Imagine if somebody said, "I'm perfect and getting more perfect, and I'm looking for someone who is perfect and willing to become ever more perfect." This would be a disastrous kind of relationship to be getting into. Forgiveness and tenderness and sympathy are based on an acceptance of our own fundamental imperfection. We are flawed creatures and need to keep our flaws in mind in order to be truly human.

There is something, frankly, frightening about perfectionism. We are angry whenever we believe that we are promised paradise and get a traffic jam, lost keys, a disappointing relationship, a less than optimal job. We are furious, and our sense of entitlement comes back to bite us. This is the danger of the age. We cease to appreciate things when we believe that life should be perfect and we can eradicate all known problems.

Why do old people love flowers? They love flowers because they're so aware of the imperfections of life that they're willing to stop by some of the smaller islands of perfection, like flowers, and appreciate them. We don't do this. If we have such a grand narrative in mind of the perfection of the species, we will not stop to appreciate the flowers.

And, last, I want to mention the importance of humour. Humour is born out of the gap between our hopes and our reality. And those who know how to laugh are also those who know how to be sympathetic to our failed hopes and our failed dreams. And we will all have them.

I think many of the worst movements in history have been born out of the minds of people who believed in perfectionism — scientists, politicians, and others who thought that we *could* straighten things out, once and for all. And this is an incredibly dangerous philosophy of life. The perfectionists amongst us are those who very often ruin and wreck the world. And true human progress is often the work of people who are much more modest, who accept their own flaws and the flaws of others, and are not attempting to make a paradise on earth.

Christianity — and I speak to you as a secular Jew — very wisely insisted that we are frail, fragile, and, all of us, broken. That's a very useful foundation. It's a conservative (and I'm not talking politically), classical starting point that is, I believe, at the root of wisdom. And, ultimately, this debate, though it seems to be a debate around science — and we have many science-y people in the room — is really a debate about wisdom and the philosophy of wisdom that you might want to adopt in your own life. Beneath the philosophy, beneath the theories of the opposing team, is a philosophy that is incredibly brittle, possibly intolerant, and cruel. It's not a liveable philosophy. At the root of humour, humanity, gentleness, and forgiveness is an acceptance that we, with our faulty walnuts, are in perfect understanding of ourselves and the world that we live in.

We must go easy on ourselves and be extremely modest. And it's this modesty that I want to sell to you, and it's on this basis that I firmly believe you should reject the motion before you. Thank you.

RUDYARD GRIFFITHS: A spirited debate is underway. Matt Ridley, you're up next for the "pro" team.

MATT RIDLEY: What a pity that we can't have a vote between who prefers bald Brits to curly haired Canadians.

Woody Allen once said, "More than at any time in history, mankind faces a crossroads. One path leads to despair and utter hopelessness. The other to total extinction. Let us pray we have the wisdom to choose wisely." And that's the way pretty well everybody talks about the future. When I was young the future was especially grim. The population explosion was unstoppable, famine was inevitable, pesticides were giving us cancer, the deserts were advancing, the oil was running out, the rainforests were doomed, acid rain, bird flu, and the hole in the ozone layer were going to make us sick. My sperm count was falling and a nuclear winter would finish us off.

You probably think I'm exaggerating. Well, here's what a bestselling book by the economist Robert Heilbroner concluded in the year I left school: "The outlook for man, I believe, is painful, difficult, perhaps desperate, and the hope that can be held out for his future prospect seems to be very slim indeed."

It was only a decade or two later that it dawned on me that every one of these threats had either been a false alarm or had been greatly exaggerated. The dreadful future was not as bad as the grownups had told me it would be. Life just kept on getting better and better for the vast majority of people.

The human lifespan has been growing at the rate of five hours a day for fifty years. The greatest measure of misery anybody can think of, child mortality, has gone down by two-thirds in that time.

Malaria mortality is down by an amazing 60 percent in fifteen years. Oil spills in the ocean are down by 90 percent since the 1970s. An object the size of a slice of bread lets you send letters, have conversations, watch movies, find your way around, take pictures, and tell hundreds of people what you had for breakfast.

And what's getting worse: traffic, obesity — problems of abundance. Here's a funny thing: most improvements are gradual, so they don't make the news. Bad news tends to come suddenly. Car crashes make the news. Decreasing child mortality doesn't. And, as Steve says, every year the average person on the planet grows wealthier, healthier, happier, cleverer, cleaner, kinder, freer, safer, more peaceful…and more equal.

More equal? Yes. Global inequality is on the way down, and fast. Why? Because people in poor countries are getting rich faster than people in rich countries. Africa is experiencing an astonishing economic miracle these days, a bit like Asia did a decade or two ago. Mozambique is 60 percent richer per capita than it was in 2008. Ethiopia's economy is growing at about 10 percent a year. The world economy has shrunk in only one year since World War

II. It was 2009, when it dipped by less than one percent, before growing 5 percent the next year. If anything, the march of prosperity is speeding up.

But my optimism about the future isn't based on extrapolating the past; it's based on *why* these things are happening. Innovation, driven by the meeting and mating of ideas to produce baby ideas, is the fuel that drives them. And, far from running out of fuel, we're only just getting started. There's an infinity of ways to recombine ideas to make new ideas. And we no longer have to rely on North Americans and Europeans to come up with them. The Internet has sped up the rate at which people can communicate and cross-fertilize their ideas.

Take vaping. In my country there are more than four million people who've pretty well given up smoking because of e-cigarettes. It's proving to be the best aid to quitting we've ever come up with. It's probably about as safe as coffee. And it was invented in China by a man named Hon Lik, who combined a bit of chemistry with a bit of electronics. Today, inventions are happening everywhere and we're benefiting from them.

But isn't all this progress coming at the expense of the environment? Well, no, often the reverse. Many environmental indicators are improving in many countries—more forests, more wildlife, cleaner air, cleaner water. Even the extinction rate is coming down compared with a hundred years ago for the creatures we know most about—birds and mammals—thanks to the efforts of conservationists. And the richer countries are, the more likely their environment is improving. The biggest environmental problems are in poor countries.

But what about population? The population growth rate of the world has halved in my lifetime, from 2 percent to one percent. And the birth rate is plummeting in Africa today. The world population quadrupled in the twentieth century, but it's not even going to double in the twenty-first. And the UN thinks it will stop growing altogether by the 2080s. Not because of war, pestilence, and famine, as gloomy old Parson Malthus feared, but because of prosperity, education, and health. There's a simple and beautiful fact about demography: when more children survive, people plan smaller families. With slowing population growth and expanding farm yields, it's getting easier and easier to feed the world. Today, it takes 68 percent less land to grow the same amount of food as fifty years ago. That means more land for nature. In theory, you can feed the world from a hydroponic farm the size of Ontario and keep the rest for wildlife. And the planet is getting greener. Satellites have recorded 14 percent more green vegetation than thirty years ago, especially in arid areas like the Sahel region of Africa.

But am I like the man who falls out of the skyscraper and as he passes the second floor shouts, "So far so good"? I don't think so. You'll probably hear the phrase "turning point" in this debate. You'll be told that this generation is the one that's going to be worse off than their parents', that they're going to die younger or see a sudden deterioration in the environment. Well, let me tell you about turning points. Every generation thinks it stands at a turning point, that the past is fine but the future is bleak.

As Lord Macaulay put it in 1830: "In every age everybody knows that up to his time progressive improvement has been taking place; nobody seems to reckon on any improvement in the next generation. We cannot absolutely prove that those are in error who say society has reached a turning point—that we have seen our best days. But so said all who came before us and with just as much apparent reason."

We filter the past for happy memories and filter the future for gloomy prognoses. It's a strange form of narcissism. We have to believe that our generation is the special one, where the turning point comes. And I'm afraid it's nonsense.

To finish, I'd like to quote Macaulay again: "On what principle is it that with nothing but improvement behind us we are to expect nothing but deterioration before us?"

RUDYARD GRIFFITHS: Thank you. Malcolm Gladwell, you're up for the "con" side.

MALCOLM GLADWELL: As I was listening to the two esteemed speakers for the proposition, Mr. Pinker and Mr. Ridley, I felt like we should come up with a more elegant term to describe them. I would suggest maybe we call them the Pollyannas. And since we, strangely, don't have a woman on stage—which is odd since it's 2015—I thought we could, more usefully, refer to them as Mr. and Mrs. Pollyanna. And I will leave it to your imagination to figure out who is Mrs. and who is Mr. Although I feel that a simple look at the contents of their scalp will give you some indication of which direction is appropriate.

In any case, as I was listening to Mr. and Mrs. Pollyanna, my mind wandered, as I'm sure many of yours did as well. I began to think of all the possible scenarios under which I would agree with them. For example, suppose we add just three words to the proposition: "Be it resolved that the best days lie ahead *for Mr. Pinker and Mr. Ridley.*" I think that's absolutely the case. Mr. Ridley is a member of the House of Lords, one of the greatest sinecures in the history of the Western world. And Mr. Pinker is on the faculty of Harvard University, which might usefully be described as America's answer to the House of Lords. It's absolutely the case that their best days lie ahead. No one is going to dispute that.

Or suppose we were debating: "Be it resolved that *Canada's* best days lie ahead." That would be absolutely true. You just had a massive upgrade at the leadership position. But, sadly, not all of us live in Canada. I live in the United States. One only has to watch five minutes of the Republican presidential debates to know that that proposition is decidedly not true for those people south of the border.

But perhaps, most seriously, if this proposition was "Be it resolved that humankind's best days *historically have lain* ahead," I think the answer is absolutely yes. And that is exactly what my two opponents have done. They have beautifully made the case that, if we go back into the past and we project forward to the seventeenth century, the eighteenth century, the nineteenth century, 1950, 1975, and we fast-forward to the present day, things have been on an upward trajectory. We can all agree that's true. But this debate is not about the past, right? It's about the future. It's about whether things get better from *this* point going forward.

The notion that the future is going to get better is hopelessly naive. And the word *better* is misplaced. What we are really facing when we look at the future is a future that's different. What do I mean by that?

I was recently at a conference and I was chatting with some people who were Internet security specialists. And they asked me, "What do you worry about? What's on your mind?" And I said, "We've done a very good job recently at dealing with the everyday low-level threats that used to comprise the world of hacking—the guy from Bulgaria who wants to steal your credit card. There are thousands of those threats. And we're doing a good job of keeping them at bay." But what terrifies us is what they called digital 9/11. Someone, some nation-state, comes in, hacks its way into the electrical infrastructure of North America, and shuts off the power for a week. Or someone goes in, simultaneously hacks into a thousand cars on the 401 and causes a massive traffic jam and car accident. Now, that might not be substantially different from the way the 401 is at the present time, but it would come as something of a shock.

I'll just give you a series of random examples. Not long ago, I was reading a paper in a political science journal and it was all about the impact of cellphone use in Africa. And it pointed out that the introduction of cellphones in Africa has had an extraordinary effect on the lives of ordinary Africans. It has permitted them to do all kinds of things they could never do in the past. But, at the same time, the introduction of cellphones has made the job of co-ordinating and executing actions and military operations by terrorist groups like ISIL (the Islamic State of Iraq and the Levant) and Boko Haram a lot easier than it was in the past.

It's also absolutely the case that our ability over the last twenty-five or fifty years to deal with what might be called ordinary climate crises has gotten an awful lot better. I don't fear, nor do most people fear famine today in the same way that we might have feared it twenty-five years ago. The threat of that kind of environmental crisis is absolutely receding because of the creation of disease-resistant or drought-resistant crops and much more effective desalinization technologies. But climate experts aren't worried about those issues. They look at the recent hurricane in Mexico, which was one of the biggest and most ferocious hurricanes ever recorded, and they say: "We are concerned about the warming trends in the world's oceans and about the possibility that one of those mega hurricanes is going to come along and deliver a blow the likes of which we've never seen." The powerful engine of human activity is what allows us to create drought- and famine-resistant crops, but it's also what is driving climate change, which is a risk of a whole other order.

Now, what do all these examples have in common? What they tell us is that, as a society, we have been engaged not in the reduction of risk but in the reconfiguration of risk. You don't have to worry about a famine every five years, but you have to worry about a mega-hurricane coming along and wiping out Miami. You don't have to worry about a guy in Romania stealing your credit card, but you have to worry about North Korea coming in and shutting off the power for two weeks. And having a cellphone in Africa means that your life is a lot easier than it would have been five or ten years ago, but it also means that the threat to you from terrorist groups is greater and more pervasive than it would have been five or ten years ago.

So what does that mean for our friends Mr. and Mrs. Pollyanna? Well, it means that everything they told you is true. It's all true. I sat there listening to the two of them, nodding my head and saying, You couldn't have spoken more truthfully and realistically and honestly about the trajectory of all these different trends. But it's only half the story. To my mind, what this debate is really about is whether the change in the *nature* of the risks that we face is something that ought to scare us. And I think the answer is obvious: it should.

RUDYARD GRIFFITHS: Excellent opening statements, gentlemen. Now, to move the debate forward, what we're going to do is have each team rebut what they've heard from the other side. We're going to hear from the "pro" team first. Mr. Pinker, we're going to put three minutes on the clock. Let's hear your rebuttal to what Alain and Malcolm have said so far.

STEVEN PINKER: I'll start with my response to Alain, which is twofold. First of all, I think Alain de Botton has shown up to the wrong debate. I don't remember ever having been invited to a debate whose resolution is that in the future people will be immortal; nor that foolishness will evaporate from the surface of the earth; nor that we will never again lose our car keys. The resolution of the debate is not that a perfect world is in our future, but rather that the best days are to come.

My second response is, "Are you serious?" Are you saying you are willing to approach a peasant in Cambodia, or Sudan, or Bangladesh, or Afghanistan and say: "Listen, I've been there. You worry about your child dying or your wife dying in childbirth; you're full of parasites; you don't have enough to eat, but trust me, it's no great shakes to live in a country like Switzerland. True, your child might not die in the first year of life, but when they're a teenager they're going to roll their eyes at you. And you may not have to live under the shadow of war and genocide, but people will still make bitchy comments. And you may not be hungry, but sometimes you'll drink wine that will have a nose that's a bit too fruity." I think the billions of people on earth would respond by saying, "Thanks, but I think I would like to find out for myself rather than take your word for it."

As for Malcolm Gladwell, it is true that in the luxury of your imagination you could imagine all kinds of catastrophes — we have no guarantee that some of the scenarios played out in Hollywood won't happen or that some hacker in Bulgaria won't shut down the electrical system. On the other hand, there's a big difference between a fantasy and likelihood. There's also a big difference between a nuisance, like stealing your credit card data, and a catastrophe. And if you have the world's cybersecurity experts from every industrialized country on earth against the teenager from Romania, I'm going to bet on the experts. Finally, if it were true that cellphones caused as much harm as the problems that they eliminated, you would see that the rate of death from warfare—can I complete the sentence?

RUDYARD GRIFFITHS: We'll let you pick up that point when we move into the moderated discussion. The three minutes has expired; we run a tight ship around here! Matt, you're up next.

MATT RIDLEY: Well, Mr. and Mrs. Cassandra did a good job of giving you the other side of the story. Seriously, Malcolm, are we to genuinely believe that a massive traffic jam caused by a teenager in Bulgaria is a catastrophe? There are far greater problems in the world. The Cassandras have told you about the problems of rich countries in the developed world.

But forget losing the grid for a few minutes because of a kid in Bulgaria who hacks into it. There are a billion people in the world who don't yet have access to electricity. And we know that when people get electricity it transforms their lives. We also know the only thing stopping people from getting power is will, resources, and that kind of thing, so there's every reason to think that we can tackle those problems.

And, as Steve was just saying, if it was really true that cellphones make war worse, we'd see it in the numbers. And we're not. Instead, cellphone access gives the average African the chance to do mobile banking, advertise his number so he can get a job, communicate with his friends relatively cheaply, and has brought wonderful improvements to people's lives. People were able to organize wars before cellphones and they still will be able to afterwards.

As for Alain, I think I heard you define poverty as a millionaire who thinks he hasn't got enough. I don't think that's poverty. I think poverty is when people really can't afford to feed themselves or to survive. And that's what I'm concerned about; I'm not looking back at the past, as Malcolm said. I specifically said I wasn't basing my optimism on that approach. I'm basing it on what we know we can do to improve people's lives—the people who really need it, who are the people in the developing world.

This world isn't perfect, definitely not. That's the whole point of optimism. Voltaire defined the word *optimist* as someone who thought the world already was perfect. Now it means something different. It means you don't think the world is perfect, but you want to improve it. And if that means that when we go to Switzerland we stop being able to appreciate flowers and we lose our sense of humour, well, maybe it's a price worth paying.

RUDYARD GRIFFITHS: Well done. We'll now get the rebuttals from the "con" team. Alain, you're up first.

ALAIN DE BOTTON: The Pollyannas are trying to get us to feel that their approach is somehow slightly risky, very modern, subtle, and interesting. It is, in fact, the conventional boosterish philosophy of the mainstream press, of capitalism, and of science. It's the message that you hear all the time. It's the new iPhone; it's the new little widget that's going to make your life better. We are surrounded by voices like those of Mr. and Mrs. Pollyanna, and there's nothing particularly fresh or interesting in what they're telling us. We hear all the time that we're heading for sunlit highland.

There are a few voices in newspapers like the *Guardian* and the *New York Times* who will say that everything is grim, but these are minority voices. The big megaphone is given to people who tell us that life is going to be perfect. This is part of what I'm fighting against, because I think it's such a dangerous and inhuman philosophy. At the dawn of Western civilization the ancient Greeks invented a form of theatre called tragedy. And the point of tragedy was to remind a city-state of its constant vulnerability and, therefore, of its need for extreme modesty in the face of the unknown.

And this is what disturbs me so profoundly about Mr. and Mrs. Pollyanna: their arrogance. They are both charming people, but their attitude is of an underlying, brittle arrogance, which I think is ultimately dangerous for us. They have an extremely materialistic view of human beings, as if the only concern that they have is the material side of life. Now, it's very easy to say, "This guy's only concerned with the problems of the rich world." But we in Canada live in the rich world. There are twenty-two countries that qualify as rich. And let's not forget them, because the whole rest of the world — the people *they're* interested in — are trying to become rich. So the problems of the rich world are the problems we need to be looking at. And they tell a very complicated story: even when the last malaria bug has been eradicated, humankind remains incredibly vulnerable to a host of challenges.

Mr. and Mrs. Pollyanna refuse to look at the real drama of what it is to be human and to take these challenges as seriously as Aeschylus, Flaubert, and Tolstoy. That is what disturbs me in the attitude of modern science and its boosterish philosophy. So I do urge you to look rather more carefully at the proposition. Thank you.

MALCOLM GLADWELL: While I was listening, I tried to make a list of statements that struck me as odd. So I thought I would just go through them, focusing on the three that struck me the most. The first was a comment by Mr. Pinker about how he was greatly cheered by the fact that the number of nuclear weapons has been reduced by 80 percent.

Forgive me for pointing this out, but it doesn't actually solve the problem. You may reduce the number of weapons by 80 percent, but all it takes is one weapon in the hands of one crazy person to blow us all up. Think of it like a person holding a gun to your head who says, "Don't worry, I've reduced the number of bullets in the chamber by 50 percent." I would not be terribly relieved by that. Perhaps Mr. Pinker would be.

My favourite Pinkerism, if I might coin a phrase, was something he said about climate change, which he glossed over very quickly. He dismissed it with the phrase, which I just love: "Economists agree it's a solvable problem." Where should I start? First of all: Economists? In what fantasy world do you imagine that economists are the first group of people you turn to for solutions to life's most complicated problems? This isn't an issue that can be solved with demand curves and by moving the interest rate up or down a basis point. Climate change is embedded in some of the most complex social, political, and economic problems of our day. It is about changing institutions and about confronting entrenched interests and the way we behave. It strikes me as typical of the kind of intellectualized fantasy world that I think our opponents are living in that they would look at this extraordinarily complex issue and turn for help to the economics department of Harvard University — or, as the case may be, the House of Lords.

And the last howler — there were too many howlers, really — from Mr. Ridley, was his notion that we could feed the world from a hydroponic farm the size of Ontario. I have only two questions for him. First of all, how much science fiction *do* you read? Second, how much hydroponic food have you eaten? Here's the big picture: it's really easy to imagine a more perfect world. It's a lot more difficult to put many of those utopian notions into practice. On this side of the aisle, we are committed to reality. And on that side, they've read just a little bit too much propaganda.

RUDYARD GRIFFITHS: What a fabulous first half of the debate. Now we're going to move into our moderated free-for-all. I have all kinds of probing and searching questions in my hand, but since the debate is flowing I'm simply going to turn it over to you, Matt. Can you respond to what you've just heard from Malcolm, particularly on this point of economics and our ability to use science to predict the future that you think we could and should inhabit?

MATT RIDLEY: The notion that we're materialists is one that needs to be disproven. It's very well to say that materialism doesn't satisfy all of your needs, but I think I would rather be well fed and miserable than hungry and miserable. So satisfying material needs does matter.

And as for Malcolm's point about climate change and climate science: all I hear from him is a counsel of despair. I don't hear him suggesting things we can do with nuclear technologies or other technologies. There are all sorts of improvements we can make. We're trying to summon the political will, we're trying to get the economics right, and we're trying to get the technologies right. We haven't succeeded yet in decarbonizing the world economy, but to think that it's completely impossible that we could even crack that problem over the next few decades is weird. I mean, we may not succeed, but it's pretty likely that we will.

MALCOLM GLADWELL: If I might beg to differ, I never said that it was impossible. Rather, I said that it was more difficult than you would lead us to believe, with all these fantastic scenarios about limitless progress. But, more than that, climate change represents a kind of threat to progress that is a different order than the threats we've seen in the past. And this goes back to my opening statement—

MATT RIDLEY: Is that really true?

MALCOLM GLADWELL: Absolutely.

MATT RIDLEY: I mean, human beings have faced some huge threats in the past—including famine and disease—and we've defeated them.

MALCOLM GLADWELL: You cannot point to a single famine that had the kind of global consequences that a consequential change in the climate of the earth will have. Name me a famine from years zero through 1750 that had the effect of changing the fundamental structure of the world's oceans. Can't do it, right? You can point to something that happened on a Scottish plane, but you can't—

MATT RIDLEY: No, but I can name a famine that happened in France in the 1690s that wiped out 15 percent of that country's population because we didn't have trade and so we couldn't get food to people in those days. But now, because of globalized agriculture, we get—

MALCOLM GLADWELL: The kind of climate change threat we're talking about is a lot greater than 15 percent of the population of France.

MATT RIDLEY: I think you're disagreeing with the Intergovernmental Panel on Climate Change, though.

MALCOLM GLADWELL: I guess I'll have to live with that devastating fact till the end of my days!

ALAIN DE BOTTON: These back and forths could get rather tedious if one team holds up a graph saying things are getting better and the other one points out, "No, but it could get worse." I think we almost have to take a step back and ask what is driving this camp to want to assert so rigorously—

MATT RIDLEY: From the data, Alain, from the facts.

ALAIN DE BOTTON: Right. But the data does not point you irrevocably to a sense that life is going to be perfect. So—

STEVEN PINKER: That's not what the debate is about.

MATT RIDLEY: Nobody said perfect. Where did this idea of perfect come from?

STEVEN PINKER: Not perfect, remember? You're at the wrong debate.

ALAIN DE BOTTON: Mr. Pinker, this is your great let-out clause. Whenever we point to a hole in your argument, you'll go: "Oh, I didn't mean *that*. Oh, of course *that* will remain a problem. It's *this* that I'm interested in." So we'll say, "What about the rate of gun use?" and you'll go, "No, no, no, I'm interested in the, you know, South African liver worm virus."

And so you constantly shift and give ground. So when I mention the high rate of mental illness you go, "Oh, no. I'm not interested in mental illness. I'm interested in extreme poverty." And then I say, "Well, what about the idea of *relative* poverty that the famous economist Richard Easterlin pointed out in the 1970s?" and you go, "No, no. I'm not interested in relative poverty. I'm only interested in extreme poverty." You keep shifting the ground, thereby making your own position slightly invalid. Because whenever we say, "No, there are real grounds for concern about the progress of humanity," you'll say, "Oh, it's not that bit of progress I'm interested in." So we have to define and stick with —

RUDYARD GRIFFITHS: Let's bring Steven in here and have you answer this charge. Are you being selective with your facts?

STEVEN PINKER: It's utterly false. I mentioned ten dimensions of human well-being. All of them have improved.

ALAIN DE BOTTON: Of course. And I could mention another thirty, but the point is that there are far more —

STEVEN PINKER: No, you can't mention another thirty.

ALAIN DE BOTTON: How many factors are there in human life? There are far more than ten.

RUDYARD GRIFFITHS: Alain, let's let Steven come in here.

STEVEN PINKER: What are the additional thirty on top of life, health, education, affluence, peace, safety, intelligence, women's empowerment?

ALAIN DE BOTTON: Do you know a famous novel, written in the nineteenth century, called *Anna Karenina*? None of the people in *Anna Karenina* suffered from your ten. Was it a happy story? No. And that tells us something very crucial, Mr. Pinker, about your narrow —

MATT RIDLEY: Did we claim we were going to abolish unhappiness?

ALAIN DE BOTTON: Matt Ridley, this is another tactical retreat that you keep putting off.

MATT RIDLEY: No, it's not. Happiness is getting better but it's not perfect —

ALAIN DE BOTTON: We define an area where your argument is not as strong and you say: "We're not interested in *that* bit. We're only interested in the liver worm." So why don't you say what you *are* interested in, stick to it, and defend it?

MATT RIDLEY: Alain, have you looked at the data on happiness? Happiness correlates with wealth between countries, within countries, and within lifetimes. It's perfectly true that you can be very wealthy and very unhappy.

STEVEN PINKER: First of all, the Easterlin Paradox has been resolved. I think you're a decade out of date. The idea that wealth does not correlate with happiness, which is the premise of the Easterlin Paradox, is wrong. Angus Deaton just won the Nobel Prize a couple of weeks ago for showing that. And as Matt said, it's a fallacy.

ALAIN DE BOTTON: Okay. But there are some people in this room who are a little unhappy about various things and they're not in the breadline. And you would go, "Well, I'm sorry, guys, but my data suggest that your happiness does not correlate with your income. And, therefore, your unhappiness is not really real because the data don't show that it's real."

STEVEN PINKER: What?

ALAIN DE BOTTON: In other words, your data leave so many anomalies —

STEVEN PINKER: What are you talking about? How do the data not show that happiness is real?

RUDYARD GRIFFITHS: Okay. Let's pause here just for a moment. I want to bring Malcolm in.

MALCOLM GLADWELL: I was just going to back up Alain, because I thought he was making a very good point with respect to the slipperiness of some of Mr. Pinker's positions.

I wanted to bring up one of the howlers that I didn't have time to go through in my rebuttal, which is on this very point. When he was talking about how we're a lot less murderous today than we were in the past, he points out that major developed countries haven't fought a war for sixty years. I think it is fair to say that when they did fight that war sixty years ago it was quite nasty, so it's of small consolation that the gap between wars is growing if the wars themselves are of terrifying ferocity. And if the wars themselves contain the possibility of the extinction of the planet, that is an extremely important point. So, highlighting the fact that there's been — wow! — sixty years since England engaged in a major war doesn't tell us much now, does it? We have to look very closely at the nature of that conflict, which is why I return to the point I was trying to make at the beginning of the debate, when I said there has been a change in the configuration of risk. Wars have gotten less frequent but more catastrophic.

RUDYARD GRIFFITHS: Matt, I want to hear from you because that's a big part of this debate. Does increased complexity equal more fragility? Are we loading up the system with things that are producing these beneficial outcomes now but could just as easily be reversed to produce calamity?

MATT RIDLEY: Let me give you an example of why that's *not* the case, using the famine example again. Nowadays, because of world trade, it's pretty well impossible to have a major global famine. It could only happen if there were simultaneous droughts in many different parts of the world. You won't see a lot of people dying in one area of famine because trade allows us to translate that specific food shortage into a general increase in prices around the whole world. So trade has reduced the risk by linking us all up. Some people say we've made the world more risky by connecting everyone. That may be true in some cases but not all of them. In a lot of other examples, it's actually enabled us to spread risk to reduce it.

RUDYARD GRIFFITHS: Does anyone want to add to this point?

MALCOLM GLADWELL: I thought I had conceded famines explicitly. I would be more interested if you actually tried to confront some of the cases where I disagreed with you than the cases where you agreed with me. That might be more effective as a debate tool.

RUDYARD GRIFFITHS: Steven, do you want to respond?

STEVEN PINKER: Malcolm, I certainly agree that economists are an inviting target, and it's always easy to get a laugh by making fun of them. But the problem of climate change is an economic problem. All the projections of the worst-case scenarios depend on calculations of economists; namely, how many people will burn how many units of fossil fuels?

MALCOLM GLADWELL: It's a problem that is defined effectively by economists. But that doesn't make it an economic problem.

STEVEN PINKER: It absolutely is an economic problem, because it all depends on how many people will burn how much carbon, how much fossil fuel —

MALCOLM GLADWELL: It's like saying that if an artist draws a still life of some apples then apples are an artist's problem. An apple is not an artist's problem. An apple is a fruit. It exists outside of the realm of artists.

STEVEN PINKER: Both the analysis of climate change and the possible solutions are economic problems. We know that we can have solar panels, but the question is, will there be enough solar panels to reduce fossil fuel use? We know that nuclear power can cut into carbon emissions, but by how much? We know that people could reduce their consumption enough to mitigate the problem, but will they? Under what kind of incentives? So it's *very much* a problem of economics.

MALCOLM GLADWELL: If I might say, this goes precisely to the point —

STEVEN PINKER: And William Nordhaus's *The Climate Casino* is the most comprehensive analysis of the chemistry, the history, the economics, and the technology of climate change. He is an economist, not at Harvard, though; he's at Yale.

MALCOLM GLADWELL: Well, I think this exchange perfectly encapsulates what Alain and I have been arguing, which is that there is something very narrow and almost precious in the way that you have chosen to look at the world.

MATT RIDLEY: "Narrow and precious." You're very good at adjectives, Malcolm. But can we have some facts?

MALCOLM GLADWELL: Let's use the example of climate change: just because economists effectively describe it does not make it an economic problem. It *is* a problem, but to successfully confront climate change will require the successful co-ordination of many different sectors of society on many different levels. To simplify the issue and suggest that it is something we can reduce to a matter of economic analysis is foolish.

STEVEN PINKER: Malcolm, it is not foolish. That's what economists do. Economics is a study of complex interactions across societies—

MALCOLM GLADWELL: You're absolutely right, which is my point. That's why—

RUDYARD GRIFFITHS: This is not a debate about economists, so I'm going to ask Alain to refocus us.

ALAIN DE BOTTON: You are approaching this debate from a scientific background and I'm coming from a humanistic background. So, for me, the history of the arts is really the description of various forms of human unhappiness, various dilemmas that humans have been in over the centuries. And that's what the history of literature, theatre, and poetry is really charting.

And I would just like to ask our learned friends how their laboratories might try and cope with some of the problems that literature has tried to address. What would you do if Hamlet walked into your lab? How would you view some of the dilemmas raised by Euripides? Or assess the levels of human unhappiness that were spotted in Kafka? Would you apply certain forms of medical intervention? How would you attack these?

MATT RIDLEY: Alain, do you think scientists are not human beings?

ALAIN DE BOTTON: You're putting that belief to the test here.

MATT RIDLEY: If you cut us do we not bleed?

ALAIN DE BOTTON: No. Well... we'll try that later! But, Matt, you're not really addressing these problems. I'm just keen to find out what answers science might have to the very serious kinds of human unhappiness that have tracked human beings throughout history.

RUDYARD GRIFFITHS: The question here is about exterior versus interior progress and the extent to which you have a feeling or a theory that progress affects not just the things that humans create but humans themselves.

MATT RIDLEY: Do I think that we are going to cure unhappiness with a pill in the next few years, so that Alain's literary heroes can be less miserable? No, I don't. But I do think that the progress of science—including the discovery of deep geological time, the vast excesses of space, and the genome of what's going on inside our cells—has enlarged the human imagination and given us even more exciting things to think about, which will influence the creation of literature and plays and things of those nature.

ALAIN DE BOTTON: So Anna Karenina is standing on the edge of the platform and you're saying, "Hang on, deep geological time is the answer for you, dear." That's your answer, Mr. and Mrs. Pollyanna?

MATT RIDLEY: Why not? Let's try it. Why don't we ask her?

ALAIN DE BOTTON: Deep geological time—fantastic. Tell that to Hamlet as well.

STEVEN PINKER: Just to remind you, Anna Karenina didn't actually exist. Neither did Hamlet. We're talking about billions of people who don't see their children die in the first year of life, who don't—

ALAIN DE BOTTON: Another classic shift of the goal post.

RUDYARD GRIFFITHS: I'll let Steven finish here.

ALAIN DE BOTTON: I understand. But Steven, please address the topic under consideration, which is the dilemmas of the psyche. Matt suggested that deep geological time would be a suitable answer.

STEVEN PINKER: If your child dies in the first year of life, that deeply concerns your psyche and it's very related to happiness. I think if billions of people do not see their children die, that's a much more relevant consideration for the human psyche—for the depths of human existence—than *Anna Karenina*.

ALAIN DE BOTTON: I just want to point out that you're doing a classic move: you're essentially saying that the problems spotted in literature are not real problems. You're like Dickens's Mrs. Jellyby; the only real issues are problems of extreme poverty.

MATT RIDLEY: I thought we were here to talk about progress, not literary theory? I'm sorry; I believe I'm in the wrong debate.

ALAIN DE BOTTON: So, in other words, extreme poverty is the only real problem on earth. Whenever I try to shift the conversation toward the issues —

MATT RIDLEY: No. *That is* the problem we're here talking about. We'll have a debate about literary theory afterwards.

ALAIN DE BOTTON: I didn't hear anything in the resolution that limited the discussion to science and material progress. I know that you're a scientist. But we are, as human beings, matter and spirit. And it behooves us on this panel to discuss both.

RUDYARD GRIFFITHS: Gentlemen, I'm going to bookend this portion of the debate because we're coming up against time and there are a few more issues I want to move through on this moderated free-for-all.

MALCOLM GLADWELL: Can I make one very quick interjection? The position of our esteemed opponents reminds me of that great Yiddish expression, "To a worm in horseradish, the world is horseradish."

MATT RIDLEY: I'm just a worm in horseradish.

RUDYARD GRIFFITHS: Matt, I want you to tell us a bit about your idea about why progress is accelerating, because that's part of the argument here: better days lie ahead, not simply because things are improving, but because of the pace at which they're getting better. What is the theory for this acceleration?

MATT RIDLEY: I wouldn't stake my argument on the fact that it's definitely accelerating, but I think there's every possibility that it could. More people are in contact today and there is a larger pool of people doing strenuous innovation work in countries other than North America and Western Europe, so the chances of life-saving innovations coming from anywhere in the world are improved. For example, this morning I read a story in the British newspapers about a baby whose cancer has been cured by gene therapy, which is a first in London. I imagine there's probably something similar happening in Japan, and in some other countries, so, all over the world we're coming up with new ideas.

But it's certainly true that improvements don't go at the same rate. If you go back fifty years, everybody thought we were going to see spectacular improvements in transport, and we haven't. However, they didn't anticipate that we were going to have such spectacular improvements in communication. So communication has accelerated much faster than we expected; transport has gone much more slowly. I think we're on the brink of a biomedical revolution that is going to be quite extraordinary. Actually, we're not on the brink; we're into it already. The most amazing things are happening in biotechnology and in the treatment of diseases, which are truly very positive for people.

RUDYARD GRIFFITHS: Malcolm, I want to hear your reaction to that, because I think a lot of people feel in their lives a sensation of acceleration, whether it be technological or innovative. Why, in your view, does that not speak to some fundamental shift that then supports the resolution?

MALCOLM GLADWELL: Well, one reason is that the very things that can create a dramatic shift in the progress of certain kinds of change can also create a parallel increase in risks. So when Matt Ridley talks about increasing global connectedness and how that leads to all kinds of positive outcomes, I believe it also leads to all kinds of negative outcomes.

Talk to epidemiologists: they will describe the threat of species extinctions, referring to human beings. The reason we are talking now about the possible extinction of human beings is precisely because we are so connected. That makes it possible for some unbelievably lethal organism or virus to spread all over the globe very, very quickly. Epidemiologists will tell you that we've come awfully close to something like that on a number of occasions quite recently because of that fact.

I want to refer back to Matt Ridley's comment about a biomedical explosion. He's absolutely correct that there have been extraordinary and unbelievably positive changes in medical technology, and in our ability to address and treat certain diseases. But let us not lose sight of the fact that when you create those kinds of new technological approaches, you create a whole series of new social and economic problems. For example, how do you pay for them? Absolutely everyone who has examined many of these new advances in medicine concedes that they come with a price tag that is, by definition, five or ten times the cost of existing therapies. You have to deal with that fact.

MATT RIDLEY: Not always, actually, Malcolm. There are things that are getting—

MALCOLM GLADWELL: One last point. All I'm saying is that the new problem you have created means that we have to temper our enthusiasm about the progress that we've made. It has created a problem even as it has solved another.

STEVEN PINKER: In other words, we must listen to economists!

Now, if you're bringing up infectious disease, there's just no comparison between the vulnerability of the human population in the past compared to the present. Matt mentioned an epidemic in France, but there's also the Black Death, which wiped out a quarter of the population of Europe. The Americas were decimated by the introduction of diseases from the Old World, as was the Old World by the introduction of syphilis from the New World.

The rate of death from infectious diseases has absolutely plunged, and there are dozens of new antibiotics in the pipeline. Of course, there are science fiction scenarios in which the proverbial Bulgarian teenager invents a superbug in his garage. But a massive and increasingly sophisticated network of expertise in molecular biology is mastering the machinery of life in a way that mitigates risks and makes them a tiny fraction of what humanity has lived with throughout its existence.

MATT RIDLEY: Earlier, one of you—I can't remember which, because you both look so alike—mocked the guinea worm, if I remember right. It's worth just mentioning that the guinea—

ALAIN DE BOTTON: I think that was just after you had mocked Aeschylus and Hamlet.

MATT RIDLEY: Could be.

ALAIN DE BOTTON: Yes, I think it was just then.

MATT RIDLEY: It's worth reminding everyone what the guinea worm does. There were 3.8 million people with guinea worms in the late 1980s. You get it from infected water. It grows down inside your leg until it comes out of a boil in your foot—

ALAIN DE BOTTON: I'm with you on the guinea worm. I will concede it.

MATT RIDLEY: You have to wrap a matchstick around the worm to draw it out, an inch at a time, over several months. Jimmy Carter is committed to its eradication, saying all we needed was better-filtered drinking water.

Last year, there were about forty cases in South Sudan—that's all that's left.

ALAIN DE BOTTON: Of course. It's a marvellous thing. It's not the case that everything is getting worse or that no progress has been made. It's the attitude toward the future that I think we're trying to put our finger on. Mr. Pinker made a charming—

MATT RIDLEY: So we're not allowed to be optimistic—

ALAIN DE BOTTON: There was a charming moment in the green room earlier today when Steven said: "Given my philosophy, wouldn't it be funny if, walking home, I happened to be bludgeoned to death by a stranger?" And we laughed and Steven laughed. And I think it comes to the heart of what we're saying. The reason it was funny—

MATT RIDLEY: I thought the green room was off limits.

ALAIN DE BOTTON: There's an old Jewish saying, "Man thinks, God laughs." In other words, it's a lesson in modesty. And Steven was, in his little throwaway joke in the green room, admitting that whatever his theories, he remains vulnerable and mortal. He might be bludgeoned to death, and his grand Pollyannish narrative could be undermined. Ultimately, this is what Malcolm and I are saying. We're not suggesting that all advancement is bad. We're simply trying to caution that your point of view has a, kind of, secularized scientific, messianic tone that can be pretty grating and can get in the way of properly accepting what life is going to be like, which is cyclical.

MATT RIDLEY: Back to the adjectives again. Alain, are you seriously suggesting that if Steve is bludgeoned to death tonight I should give up my view of the world getting better?

ALAIN DE BOTTON: If you put a drape over Steven's body and saw him off to the morgue, and then reassured yourself with some statistics that it was actually very unlikely that your dear friend be bludgeoned to death, and went home to bed and managed to fall asleep, you would be proving our point of your inhumanity.

MATT RIDLEY: No. I would go to his funeral. I would be very unhappy. But I wouldn't suddenly turn around and say that proves the whole world is getting worse.

MALCOLM GLADWELL: We would settle for five minutes of introspection —

STEVEN PINKER: I think I should have a say in this.

RUDYARD GRIFFITHS: Yes, absolutely. We might organize some extra security on the way home tonight.

We're going to wrap up the free-for-all with final remarks from Malcolm. And then, Steven, we'll give you the last word. Malcolm, go ahead. Did you have something to say?

MALCOLM GLADWELL: Well, I was wondering whether *you* had a question, but—

RUDYARD GRIFFITHS: No, no. I wanted to just bookend this discussion about Steve's mortality and go on to our closing remarks.

MALCOLM GLADWELL: Many things occurred to me as I listened to the two of them. I suppose I can recount my thoughts at this time.

The first was that I wish I had the kind of cheerful self-confidence that those two have, so that whenever I were to imagine a worst-case scenario I could discuss it as a science-fiction fantasy. What a wonderful way to banish all unfortunate thoughts. I wish that had occurred to me as a teenager when I was at my most troubled and angst-ridden. I would have had a much happier adolescence.

My other point was—

ALAIN DE BOTTON: They're going to bring up the guinea worm. Any vulnerability, and they're going to bring up the guinea worm.

MALCOLM GLADWELL: And my other point, in listening to Matt Ridley in that quite hilarious discussion about the hypothetical bludgeoning of our dear friend Mr. Pinker, echoes Alain. We're asking for just a little moment of introspection, a little understanding that these questions cannot be resolved entirely through a simple appeal to statistics and to what ran in *Nature* or the journal *Science* last week. We would like them to step outside of their very narrowly constrained, scientific universe and just consider these problems in the light of their full complexity.

RUDYARD GRIFFITHS: Excellent. We're going to give the last word for the moderated portion to you, Steven.

STEVEN PINKER: Well, since a lot of this debate seems to hinge on my imminent bludgeoning, I'm willing to accept those odds! I think that if I get bludgeoned to death tonight, Matt will concede. But if I can tweet tomorrow morning that rumours of my death are greatly exaggerated, then I would maintain that our side should win because the chances of any of us being bludgeoned to death are extremely small and a fraction of what they were several centuries ago.

In terms of simplicity versus complexity, Malcolm, are you suggesting that a scientific approach to progress is simplistic? That instead we should look to Aeschylus or we should look to science fiction? I would maintain that if you want to understand the world and the way it's going, which scenarios are likely and, most important, how to deal with them, science is the sophisticated way to address these problems, not the simplistic one. If you want to know what we should do to continue the trajectory to reducing disease, hunger, and climate change; to increasing lifespan; to getting kids to school; then yes, science is where you should look. It is not the simplistic way to deal with these problems or to analyze them. I don't believe fiction is the appropriate way to figure out how to deal with the very serious challenges that we have. We should deal with them through science.

RUDYARD GRIFFITHS: Very good. Let's move on now. This is an excellent free-for-all. Because we've just heard from Malcolm and Steven, I'm going to change the order of closing statements slightly. So, Alain, we're going to have you go first.

ALAIN DE BOTTON: I'm glad that Steven really showed his true colours. I didn't think he would just come out with something so crass.

I want to freeze the moment. One of our great scientists has said that literature is not real, that it's made up. In other words, he's suggesting the work of the imagination is something that has no validity. He is arguing that science has the answers and the humanities have none. This is what I was terrified of, and Steven has reassured me that I was right to be afraid. It's precisely this attitude that is so dangerous in scientists. Historically, the great scientists have known their limitations and have worked with the humanities to understand the complexities of the human mind.

What you have in front of you, ladies and gentlemen, is a new kind of scientist who is so cocksure of what he and his lab can do that he has discarded two thousand years of insights from the humanities, from religion, and from anything that lies outside of the scientific method. And this is highly reductive and highly dangerous.

In the past, people who were very religious dreamt of

a New Jerusalem, a new dawn, when all problems would be done away with through the light of reason. The United States is just across the border, and it was built upon the idea of constructing heavenly Jerusalem here on earth through the use of religion. What these two gentlemen represent is a secularized, scientific version of that New Jerusalem. It was dangerous then and it's dangerous now because it breeds millennial fantasies of perfectionism, which are very dangerous.

In Canada, in Europe, and in other parts of the world, we remember older, more complex kinds of heritage, where we accept that human nature cannot be perfect, and that the best way of improving our laws, our societies, and our relationships with one another is not through statistics or the assumption that science can provide all the answers.

Believe me, I am a firm believer in the wonders of science. Like everybody, I'm deeply supportive of those brave researchers who have wiped out all sorts of diseases in Africa. But don't, ladies and gentlemen, allow this to sway you in assessing the motion. Because you can very much feel proud of what scientists have done without wishing, as this team would like us to do, to disregard all the complexities of the psyche, or supernatural inclinations that we call the soul. Those struggles continue, and we have philosophy, art, and other disciplines to deal with them, which stand shoulder to shoulder with the sciences in the hope of making life not necessarily more perfect, but sometimes less painful. This is what I am arguing for: a more humane, realistic version of the meaning of life. Thank you.

MATT RIDLEY: "It is not the man who hopes when others despair who is regarded by a large class of persons as a sage," said John Stuart Mill, "but the man who despairs when others hope." In other words, we've always thought Cassandras were wise and Pollyannas were foolish. But history teaches us that this is the wrong way around. Cassandras were nearly always wrong and Pollyannas have rarely been cheery enough, given our history.

The apocaholics of my youth were wrong to tell me that the world, and the future, was grim. And they were wrong to teach me a counsel of despair about it. But don't go away with the idea that optimists like Steve and I think the world is perfect. I have no idea where Alain got that idea. Of course we don't think that. We think quite the reverse. We think this world is a "vale of tears," a "slough of despond" compared with what it could be—and will be in the future—if we do the right things.

I'm not an optimist by temperament, but by evidence. That's what changes my mind. We're not saying, "Don't worry; be happy." We're saying, "Don't despair; be ambitious." We're not saying everything is going to be okay. There's going to be war and pain and misery in the future, but there was even more in the past. Speaking of the past, I'll give Alain a little bit of ancient Greece. Hesiod lived in the golden age of Greece. And even he complained that things weren't what they used to be.

We've hardly started gathering the harvest of innovation that can improve people's lives in the future and heal the planet we live on. And that's what makes Steve and me different from Dr. Pangloss in Voltaire's *Candide*.

Pangloss explains to Candide that the death of 70,000 people in the Lisbon earthquake must be for the best, because God made the world and he's not able to make an imperfect world, so they must have been bad people. Voltaire was actually teasing the *théodicée* of Leibniz, and Pierre Louis Maupertuis, maybe because his mistress was sleeping with Maupertuis. Today, we would call Pangloss a pessimist, somebody who thinks that we really can't improve our lot, and that the world is as perfect as it could be. Progress has been real; progress is real; progress has been good for the great majority of people. Progress has been particularly good for poor people. And there's no reason to think that it's suddenly going to stop now just because we're not thinking enough about our soul or our psyche. There's every reason to think that the future is going to be bright. And I think you should vote for the motion if you think that.

MALCOLM GLADWELL: I want to talk about something that we haven't spoken nearly enough about this evening, which is kind of astonishing, but maybe it's because the two esteemed gentlemen on the other side of this proposition have essentially spent the entire time with their hands over their ears and eyes saying, "La, la, la, la, la, la, la." What I'm talking about, of course, is nuclear war. And the story that always stays with me and I think bears repeating this evening is the story of what's known as the Petrov incident. September 26, 1983: relations between the United States and the Soviet Union were at an all-time low. Korean Air Lines Flight 007 had just been shot down by the Soviets. We were on the brink, as close as we have come, perhaps, to nuclear war in quite some time. The hawks were all lined up in Washington; Yuri Andropov in Moscow believed America was on the verge of a nuclear first strike.

And at that point—at the height of paranoia—a lieutenant colonel named Stanislav Petrov in the Soviet Air Defence Forces looked at his computer screen and saw a report of an incoming nuclear missile from the United States that could blow up the Soviet Union. And he knew what protocol required him to do: if America launched a first strike, he had to retaliate in full force.

So what did he do? Well, he decided that it was a computer malfunction and he didn't push the button. And his hesitation is the real reason that we're all here today.

The lesson of the story is obvious. This couldn't have happened in the fifteenth century, as miserable as that century was. It couldn't have happened in the sixteenth, seventeenth, eighteenth, nineteenth, or the twentieth, up until 1940 or so. And this threat is a distinctly modern creation.

The notion that a computer malfunction could lead to us all being blown up—the threat of that—is as real today as it was thirty years ago in the Petrov incident. And so I return to the point that I made at the beginning of this debate. We have done extraordinary things over the last hundred, two hundred, three hundred years in reducing our interpersonal risks and in making progress in the everyday ways we live our lives.

Mr. Pinker gave you ten areas in which we have

made that kind of progress. Matt Ridley gave you more. Everything they said is true. But it's beside the point. While we have reduced those interpersonal risks, we have increased our existential risks. And by voting for the resolution, you would have to believe that these trade-offs leave us better off. And they do not.

RUDYARD GRIFFITHS: Steven.

STEVEN PINKER: Everyone knows that the human mind is vulnerable to illusions, biases, and fallacies. Several of these mental bugs fool us into believing that the world is in decline or in existential danger and always has been.

First, we're overly impressed by memorable images. And that's what the world gives us. If it bleeds, it leads. That's why we fear shark attacks and plane hijackings when what we should fear is falling down the stairs and texting while driving. Now, smart phones have made billions of people into on-the-scene reporters and the world seems to contain more shootings and explosions than ever.

Second, we confuse changes in ourselves with changes in the times. As we get older, we become more aware of follies and dangers that have been there all along. So every generation becomes nostalgic for the good old days.

Third, everyone is a social critic. As Hobbes observed, "Competition of praise inclineth to a reverence of antiquity. For men contend with the living, not with the dead." Recently, the epidemiologist Hans Rosling gave a thousand people a series of multiple-choice questions on population, literacy, life expectancy, and poverty. He noted, "If for each question I wrote the alternatives on bananas and asked chimpanzees in the zoo to pick the right answers, they would have done better than the respondents." The reason was that the respondents consistently picked answers that were too pessimistic. I got the same result in a survey about violence in the present and in the past. And this refutes the claim by Mr. Botton that people's default position is one toward optimism. The facts are exactly the opposite.

Ladies and gentlemen, you can do better than a chimpanzee. The cure for cognitive fallacies is data, and the trend lines are unequivocal. On average, people are living longer, healthier, richer, safer, freer, more literate, and more peaceful lives. While past performance is no guarantee of future returns, the world is not Wall Street. We are unlikely to wake up one morning and face a world with smallpox, slave auctions, or surgery without anaesthetics.

To be sure, the world faces formidable challenges. And that brings me to my final point. Optimism is a self-fulfilling prophecy; so is pessimism. The progress we enjoy is not the result of some mysterious historical dialectic or law of inevitable progress or arc bending toward justice. It is the result of people spotting problems, including fragility and nuclear proliferation, and instead of moaning that we're all doomed, applying their ingenuity and their efforts to solving them.

A recent survey showed that people who believe that our way of life will end in a century also endorsed the statement, "The world's future looks grim, so we have to focus on looking after ourselves." Ladies and gentlemen, don't be among them. It's irresponsible enough to be a fatalist when the objective indicators say the world is getting worse; all the more so when they say the world is getting better.

RUDYARD GRIFFITHS: Thank you, gentlemen. Bravo. Well, ladies and gentlemen, this has been a big, meaty debate. I just want to thank the four debaters: you came here prepared, took each other on, and we're the better for it. Thank you again for a fabulous debate. Well done — it was a terrific contest.

Summary: The pre-debate vote was 71 percent in favour of the resolution, 29 percent against it. The final vote showed 73 percent in favour of the motion and 27 percent against. Given that more of the voters shifted to the team supporting the resolution, the victory goes to Steven Pinker and Matt Ridley.

Pre-Debate Interviews

MALCOLM GLADWELL IN CONVERSATION
WITH RUDYARD GRIFFITHS

RUDYARD GRIFFITHS: Welcome to our pre-debate interviews for the Munk Debate on Progress. We'll be debating the resolution, "Be it resolved: humankind's best days
lie ahead." Arguing against the motion will be Malcolm
Gladwell, internationally bestselling author and staff
writer at the *New Yorker*. Malcolm, let's start by having
you unpack some of the key arguments that you'll be
putting on the table tonight. I think you may be running a little bit against the current of our culture right
now, where there's so much excitement around technology and some of the big global forces that are revealing
themselves.

MALCOLM GLADWELL: Yeah, and I think it's important to
distinguish that we can make progress in individual areas
without necessarily being better off as a whole, right? So,
I imagine the other side is going to provide the specific

incidences where we genuinely are better off than we were in the past, and where every expectation is we will continue to improve in the coming years. And I will concede all that to them—they're absolutely right. I can sit here and simply list things that I think will get better in the future, but that doesn't even begin to answer the question of whether we will be better off as a whole in five, ten, or twenty-five years from now.

I'm more interested in questions like, When you make progress against an individual problem do you create additional problems? What are those additional problems, and are those new problems bigger, the same as, or less than the problems you've just solved?

RUDYARD GRIFFITHS: Atomic weapons would be a great example. If we want to light up cities we can blow up cities.

MALCOLM GLADWELL: Exactly.

RUDYARD GRIFFITHS: We invent nanotechnology, weaponize it, and it wipes us out. Is it that kind of dual use in a lot of technology that worries you?

MALCOLM GLADWELL: As technology and some other areas progress, the size and the catastrophic nature of the flip-side grow exponentially. So for every huge leap you make in one area, you also create a huge leap in, for example, the ability of humankind to destroy itself.

I think it's important for us, for my side, to not come

across as pessimists. We can be the furthest thing from it. And my position is simply that the world is different, and as we go forward, I don't know whether we're better off or worse off. It strikes me that the other side's incumbent to prove the proposition. I simply have to shrug, and if the audience agrees that we ought to shrug, we win.

RUDYARD GRIFFITHS: A very French debate. Tell me about human progress in relation to morality. There is an argument out there that we're not only improving the external world, but that we're somehow improving ourselves. Steven Pinker is one of its key proponents.

MALCOLM GLADWELL: Pinker's argument, which he expresses in his book *The Better Angels of Our Nature*, is at once commonplace and also utterly beside the point. Are we better as a people than we were when we, as he put it in his pre-quote, "threw virgins into volcanoes and cut people's hands off for stealing cabbage?" Absolutely, but nobody ever said that we weren't.

It strikes me as being so true that it's almost a kind of cliché. And it does not resolve the question of whether, in the present day, the spectacular and idiosyncratic evil of a handful of individuals can make a huge difference in our life. So, 99.9 percent of us might be better off now than we were previously, but the remaining .1 percent of people can make the rest of our lives really, really miserable. It only took one Stalin to wipe out twenty million people in the Soviet Union. Even if the rest of the people living in the Soviet Union were angels, it doesn't change the fact that they suffered a very grim fate at the hands of one dictator. Ultimately, I don't think this argument is terribly useful in making sense of the world.

RUDYARD GRIFFITHS: We seem to have an obsession with progress, from the earliest origins of Christianity through the modern era, and now to our fascination with artificial intelligence. Do we use progress as a compensatory strategy to help us reconcile certain expectations of ourselves? From where do you think this fixation emerges?

MALCOLM GLADWELL: It's a very useful fiction, but it's also exceedingly beneficial when it's true. Let me give you an example. I'm a big runner. Runners like to believe in progress because every year we see world records fall, right? And because we see these records fall we participate in a sort of grand illusion that somehow the running community is making one small step forward every year, that we are doing better than our predecessors. But does a race become any more exciting just because a runner finishes the 100 metres in 9.5 seconds as opposed to 9.51? Does it make the winner any more satisfied? Does it make the crowd any more excited? It doesn't change any of the material, meaningful facts. It might just mean that someone tweaked the kind of spike in their shoe or that the surface of the track was a little bit different. I think there's a point at which we fetishize this notion of progress even when it makes no meaningful difference to who we are.

RUDYARD GRIFFITHS: It's interesting you mention running. Do you think progress is linked to a collective and personal preoccupation with winning? We like to win and we don't like to lose.

MALCOLM GLADWELL: I like that argument because it makes our opponents seem incredibly shallow and callous, which I think is a wholly appropriate reading of their motivation. It makes me and Alain look kind of highly evolved, which is probably true.

RUDYARD GRIFFITHS: Your side of the debate is progressing; their side is regressing. Malcolm Gladwell, thank you for coming to Toronto, and thank you for being part of this debate.

MALCOLM GLADWELL: Very well.

ALAIN DE BOTTON IN CONVERSATION
WITH RUDYARD GRIFFITHS

RUDYARD GRIFFITHS: I'm joined by Alain de Botton, the author, philosopher, and part of the team, with Malcolm Gladwell, who'll be arguing against tonight's resolution, "Be it resolved: humankind's best days lie ahead." Alain, great to have you in Toronto.

ALAIN DE BOTTON: Thank you so much.

RUDYARD GRIFFITHS: Give us a taste of how you're going to approach this debate tonight. I would think you might have a concern that the social consensus out there could be on Steven Pinker and Matt Ridley's side, just in terms of how people think about their future and what's in store.

ALAIN DE BOTTON: We live in a world that has two very powerful optimistic drivers behind it: business and science. We live in a commercial society, and if you want to sell

anyone anything you've got to sound positive about yourself, your future, and your prospects, so the atmosphere
we live in is incredibly cheerful. Science dazzles us and
constantly promises us enormous changes. Despite these
two very powerful forces, I want to argue that there are
stubborn and cyclical obstacles to fulfillment and to a
properly happy life.

When people think about what's wrong with the world
and suggest ways to improve it, they cite things like education. They say we don't know enough and that if we could
only get the education system right then things will be
okay. But it's never ending. People say that if we only we
get the economy growing, and end poverty, then things will
work out. And then they'll suggest that ending war and
conflict is the solution. And there's a lot of hope around
medicine, that we will be able to end a lot of suffering if we
manage to get medical advances to a certain stage.

I'm Swiss, and when you come from Switzerland you
come from a country that has solved all of those problems as best as humankind can. Switzerland is maybe five
hundred years ahead in development compared to places
like Uganda, or Liberia. The bad news is that there are
still a lot of problems there, and I'm interested in these
so-called first-world problems. That's normally seen as
a pejorative, an insult, like they don't really exist, but
they do, and they have a striking ability to mar existence.
A lot of my focus tonight is going to be on what these
first-world problems are, how stubborn they are, and how
much they impede some of our grander visions of making
life on earth perfect.

RUDYARD GRIFFITHS: Can you give us an example or two of a first-world problem that is actually something serious and substantive and isn't just the result of being at the top of the pyramid?

ALAIN DE BOTTON: One of the real tragedies of economic development is that we thought that by increasing people's material conditions we would improve human happiness. However, an economist named Richard Easterlin did a famous study on the relationship between income and happiness over forty years ago. He argued that even if a society is by all accounts extremely rich, the desire for more, envy of others, and a sense of status and anxiety continues. You don't end financial dissatisfaction merely by enriching a people. Comparativeness is a societal endemic, and envy, jealousy, and a sense of inadequacy exist among billionaires. There's enough pause for thought.

That's just one example. Even if we made the whole world into billionaires there would be economic quirks, unhappiness, and grief. I'm not saying we shouldn't pursue this — of course we should try — but don't expect that everything will be perfect simply because everyone enjoys the income level of, say, a Swiss dentist, which may happen on the planet in five hundred years.

RUDYARD GRIFFITHS: Tell us a little bit about what you think you're going to hear from your opponents tonight, which will probably be quite data driven. They're likely going to look at the eradication of diseases, the reduction in acute poverty throughout the developing world, and the extent to which life expectancy has generally risen across the globe. If you look at a data frame from over the last hundred years or so, those lines look quite impressive and move up the chart nicely. How do you respond to that idea?

ALAIN DE BOTTON: What our optimistic friends are missing is just how perverse the human mind is. No one will want to deny the upward curve that these nice folks are pointing to, but we have to balance such things as the ability to increase educational standards and to massively increase the knowledge that people have at their fingertips with the persistence of idiocy and ignorance despite education. The great dream of the Enlightenment was that through education, people would abandon prejudice, rotten ideas, and bad impulses, and that these things would melt away like fog on a sunny day under the light of reason. It doesn't happen. And we've seen conflict after conflict in educated populations, so education is not a panacea.

Medicine is similar to education in this regard. For all the advances in medicine, there's one thing that's staring all of us in the face, which is our own extinction. We are each facing our own miniature Armageddon, our own apocalypse. Of course it's great to have advances in cardiac knowledge, in cancer treatments, and the like, but we haven't and won't ever resolve these issues.

I'm optimistic for a species that is not *Homo sapiens.* I believe that it is possible that maybe in a thousand years' time we will create a species that doesn't die, which is properly able to use knowledge, and is happy, and inherently non-aggressive. But it won't be *Homo sapiens*; it will be another species. I could be optimistic but just not on behalf of humanity. There's a better-designed *Homo* that might be coming some point in five hundred years. It's not us.

RUDYARD GRIFFITHS: It seems like part of your argument is about the qualitative as opposed to the quantitative, and the extent to which we are obsessed with the quantitative in the modern world. We focus on the pure empirical measurement of ourselves against others, of our society as opposed to other nations, whereas you think the qualitative dimensions of our lives, our inner selves, is still a field of poverty in a way.

ALAIN DE BOTTON: Lots of things slip through the net of statistics. You cannot simply draw a line in health, income, social status, and say people who are situated above this line will be content, because so much has got to do with this tricky internal feature called expectations.

One of the things to bear in mind is that expecting that the world won't improve does not mean that you are a cheerless, depressive soul. Indeed, what I want to recommend to the audience is that the capacity to stare the grim facts of existence properly in the face endows one with a certain buoyancy and capacity to deal with problems. Part of the brittleness and embitterment of the modern world has got to do with expectations that were never met and a sense of entitlement that was betrayed. And here it's very wise to take a leaf from the ancient stoics, who very much recommended a philosophical approach to life where you expect that a lot of things may go wrong, but that you can survive them, and if you don't your end may be quick.

And this sounds like a very alien philosophy, particularly in North America, but I think it's the best hope we have. I don't think that people who are pessimistic have necessarily given up on human progress. One can be a great believer in human advances while keeping the dark facts very much in view. So, I want to rescue the anti-progress group from a kind of vision of darkness and depression. We're not depressive. We're cheerful realists.

RUDYARD GRIFFITHS: Right. What do you make of the fact that as humans we've been obsessed with progress now for almost two millennia? We've had a Christian theory of progress, and theories in the twentieth century built around certain internal dynamics of history. Tonight, we'll no doubt hear a story from Matt Ridley about how social evolution is the new springboard for progress. What does this mean? Why have we been so obsessed with this idea for so long?

ALAIN DE BOTTON: Well, Rudyard, you put your finger on something very important in mentioning religion and connecting that to science. Science is a secularized version of the Christian narrative of the perfection of humanity, and this has been something that has haunted the Western imagination since the year dot. We can imagine perfection, which has something to do with the fact that we have very powerful tools of imagination and it isn't hard for us to simply subtract everything that's negative from our own characters and our own societies and imagine heaven on earth, the New Jerusalem.

Our scientific friends often like to pretend that they don't speak in a language that points to a scientific version of the New Jerusalem. I think it's a beautiful dream and it's very important to have these ideals because they provide a roadmap, a sense of direction, and we're pointing toward the New Jerusalem. We'll never get there because human beings are endowed with a brain, which I like to refer to as the faulty walnut. We have faulty walnuts at the top of our spinal columns, and these faulty walnuts misfire: they're aggressive, they don't remember what they should, and they're endowed with all sorts of very unhelpful drives. We've tried to create civilization, which should correct some of the worse impulses and selfishness of the faulty walnut, and we have. Civilization is a kind of superior brain that looks after the little brains that are all flawed.

But even so, we won't entirely eradicate the inbuilt flaws. Artificial intelligence is the current dream, that we will make a perfect machine-made human. I happen to be very interested in the possibilities of artificial intelligence. But it's at the point of the most advanced kind of artificial intelligence that we actually leave *Homo sapiens* behind. So it may happen, but it won't be the human race that will have got there; we will have become something else and someone else.

RUDYARD GRIFFITHS: Wow. Alain de Botton, thank you for coming to Toronto. Fascinating argument. We look forward to the full show tonight on the main stage.

ALAIN DE BOTTON: Thank you so much.

MATT RIDLEY IN CONVERSATION
WITH RUDYARD GRIFFITHS

RUDYARD GRIFFITHS: Welcome back to our pre-debate interviews for tonight's Munk Debate on Progress. I'm joined right now by Matt Ridley, the internationally bestselling author, London *Times* columnist, and member of the British House of Lords. Matt, great to have you here.

MATT RIDLEY: Lovely to be here.

RUDYARD GRIFFITHS: Give us a sense of some of the key arguments that you're going to put forward tonight. I think you've maybe got an audience that's receptive to your and Pinker's ideas, given our enthusiasm for technology and our sense of optimism about Toronto right now.

MATT RIDLEY: Canada's certainly a good country to talk about in relation to how positive progress has been. Obviously it's a wealthy country, and it's also healthy and happy. But I'm going to talk about the whole world being that way, because over the last fifty years we've seen the most extraordinary progress in terms of human living standards. Only 10 percent of people live in extreme poverty now. It's still too many but it's an amazing change. Overall, people are wealthier, healthier, happier, cleverer, cleaner, kind of freer, more peaceful, and more equal, even. People are getting rich in poor countries faster than people in rich countries, so that's bringing global equality.

Not everything's going the right way, but the things that are going in the wrong direction tend to be less important than the things that are going in the right direction.

RUDYARD GRIFFITHS: If we went to a period in antiquity and chose fifty years, let's say, of record Roman green harvest, you'd see rising life expectancy, lower bouts of pestilence and disease, and probably a sense of progress. So what is different from this era of progress that we're experiencing now than similar eras in the past?

MATT RIDLEY: The main difference is that this one is global. In the past you've seen prosperity in islands, like the Roman Empire, or the Chinese Empire, or India, but it was never spread around the world as a whole. And the other difference is that we have got technologies now that mean that it's pretty well impossible to turn the clock back on most of these things, so it would be very, very difficult to unwind the Internet, as one example, or the technologies that have enabled us to feed the world and to get rid of disease and so on, because you could reinvent them in a flash if you needed to. So that's one big difference. And when you think about it, innovation is really what's driving progress. And because of the Internet, people can cross-fertilize their ideas quicker than ever, and that's bound to lead to solutions to whatever problems we encounter.

Now, it's not guaranteed that the future is better than the past. I completely agree with that: one should never say that it is, because an asteroid or an organic explosion or a particular kind of nuclear war or even a particular outbreak of irrationality in the human race could derail it, but I think there's more probability that it will turn out good than bad.

RUDYARD GRIFFITHS: Talk to us a bit about the social evolution of ideas. You think this is really kind of critical to why we should look at progress in new and different ways now than we did in the past.

MATT RIDLEY: Most of the good news is gradual and most of the bad news is sudden. That's why television news is dominated by bad things and good things go unreported. That's why after a century dominated by world wars and genocides we ended up better off than we were before, because the gradual goes unreported.

Gradual change is a bottom-up phenomenon. It's an evolutionary thing that happens in society: two different ideas meet, mate, and produce a baby idea that improves people's lives, and that gradually spreads through society.

Take, for example, the Internet or the mobile phone revolution. They come about in an automatic way without anybody ordaining them, and yet they're producing what's called spontaneous order. In other words, they're producing fits between form and function in society that are not intended by anybody but nonetheless emerge.

RUDYARD GRIFFITHS: And you think that because this is happening on a global scale it's both new and bodes well for these forces accelerating?

MATT RIDLEY: We have more possibilities to improve the world today than we had fifty years ago. So, I think, given the number of people working on science and technology, the number of technologies they've got at their disposal, the amount of knowledge they've amassed, there's every reason to think that we're more likely to be able to solve a problem today than we were fifty years ago, and yet we did pretty well then. So that's one reason to look hopeful.

As Thomas Babington Macaulay said a long time ago, in 1830: On what principle is it that with nothing but improvement behind us we are to expect nothing but deterioration before us?

RUDYARD GRIFFITHS: Very good. I think your opponents tonight will try to paint a picture that you're confusing the phenomenon of progress with the phenomenon of change and difference. And yes, ten years ago we didn't take selfies of each other on our cellphones, but just because we're doing that doesn't mean that we've progressed from where we were a decade ago. What do you see as the nuances between progress and difference?

MATT RIDLEY: As one example, I'd say it's pretty hard to go to a mother who's just lost a child to malaria in Africa and say ridding the world of that mortality would not be progress. Global malaria mortality has gone down by 60 percent in the last fifteen years alone, which is quite an extraordinary rate of change. That change has happened over time because of bed nets impregnated with insecticides and things, pretty low tech. So, it's not about mobile phones or computers, necessarily; it's just ordinary old human ingenuity.

And when you look at what's happening to poor people and the improvement in their lot in the world, it's pretty hard not to describe what's happening as progress.

RUDYARD GRIFFITHS: Another line of attack might be that we live in an increasingly interconnected world, and that unintended consequences as a result of mistakes and actions made in, you know, mortgage markets in southern Florida suddenly create global financial contagion, such as the 2008 financial crisis. Do you see this interconnectedness as a danger? The more our complex relationships evolve the greater the potential becomes for larger collapse?

MATT RIDLEY: Well, in some ways I think it's the opposite. The greater connectivity of the world actually makes us less vulnerable to collapses. Let me give you the example of world trade in food. Back in the seventeenth century it was possible for 15 percent of the entire population of France to die because of two bad harvests in a row, while next door in Britain there were decent harvests and people survived. And that's because there was very little trade in food: it was very expensive to move food around and it wasn't possible to feed the starving.

Today, because we have a global food trade, you'll never get a global famine at the same level as you had in the past. A failed harvest in one region might cause a slight uptick in the global price of wheat, but you'll never get a failed harvest in Australia, Siberia, Canada, and Argentina all at the same time. And we've seen famine absolutely retreat in the most extraordinary way in the last twenty to thirty years.

RUDYARD GRIFFITHS: Is there some event or idea that would make you question your thesis?

MATT RIDLEY: There are things I worry about — trends that are going in the wrong direction, and if they get too strong could derail progress and improvement.

RUDYARD GRIFFITHS: Like what?

MATT RIDLEY: Superstition. There are more and more people in the world being brought up with fundamentalist versions of religion. I'm not talking about any specific religion here; I'm talking about all religions. But the people who are in the most fundamentalist ends of the various religions are having more babies, literally.

Now, if children follow their parents, which luckily they often don't, that could lead to a huge increase in the number of people who are prepared to be extremist and fundamentalist about their views. That wasn't a problem in the old days because everybody tried to have as many kids as possible, and so they would be swamped by the children of the reasonable people, but I'm slightly worried that unreasonable people are having more children than reasonable people, if I can use that expression without offending anybody.

RUDYARD GRIFFITHS: You've not offended anyone. Matt Ridley, thank you for coming to Toronto. It's great to have you here.

MATT RIDLEY: Thank you, Rudyard.

STEVEN PINKER IN CONVERSATION
WITH RUDYARD GRIFFITHS

RUDYARD GRIFFITHS: Welcome back to our pre-debate interviews for the Munk Debate on Progress. We will be debating the resolution, "Be it resolved: humankind's best days lie ahead." I'm extremely pleased to have Steven Pinker with me—the celebrated American scientist, psychologist, and bestselling author. Steven, great to have you here in Toronto.

STEVEN PINKER: Thank you.

RUDYARD GRIFFITHS: Give us a sense of how you're going to frame tonight's debate. What's the core argument that you want to communicate to the 3,000 people who will be gathering here at Roy Thomson Hall?

STEVEN PINKER: Whether humankind's best days are ahead should not be a matter of opinion, attitude, or mood, but of facts. And if you look at the data, every indicator of human well-being is on the rise. We're living longer; we are getting sick less often; we are richer; we are more likely to live in democracies; we're more likely to be at peace; we're smarter; we're better educated. And when I say "we," I don't just mean we in charmed places like Canada, but globally, and that is what the data says.

Moreover, the processes that have pushed us in those directions are almost certain to continue. They consist of innovation and new technology and ideas — they aren't subject to chaotic booms and crashes like the stock market, so you're unlikely to wake up tomorrow to a world in which you have to have surgery without anaesthetics or see your kids have less education than you had. These are all cumulative directional forces, and there's no reason that they will evaporate overnight.

RUDYARD GRIFFITHS: Give us a sense of what you feel is driving those directional forces. Matt Ridley, your partner tonight, is a proponent of social evolution, which he thinks is a key factor in understanding why progress, loosely defined, is accelerating. What are the nuances of this aspect of the argument?

STEVEN PINKER: I'm sympathetic to Matt Ridley's argument. I think we're a smart species because of language. Thanks to the written word, we can accumulate the fruits of our trial and error, our strokes of genius, our lucky accidents, and record what works, jettison what doesn't work, and know there'll be lots of ups and downs. But overall, if you set your mind to solving problems and you remember the solutions that work in general, things are going to get better.

RUDYARD GRIFFITHS: Do you think interconnectedness is a vulnerability? Is our technological, sociological, and economical complexity an Achilles heel to the methods and mechanisms of progress that you see as happening right now?

STEVEN PINKER: The thing about rationality is that it can always step back and observe its own limitations and set that as a goal, as a problem to be solved, so the complexity of our systems in turn sets up the goal of how do we make them robust to perturbations that might be chaotic. How do we simplify where we can and how do we develop artificial intelligence to deal with some of the complexities? So the very fact that you have pinpointed complexity as a problem means that our best minds will set the goal of dealing with it.

RUDYARD GRIFFITHS: People often think of the financial crisis as a recent example of complexity. They think of a system that was thought to be benign, and which suddenly had this somewhat unexpected and shuddering near-collapse. Understandably, a lot of people might feel that kind of lessens their belief in progress and the degree to which complex systems can, day in and day out, produce positive outcomes. How would you react to that?

STEVEN PINKER: The global financial crisis only caused the world's GDP growth to halt for one year, and then it just kept going up. And it was barely felt in Asia and Africa. So the idea that things get better doesn't mean that every single day it will get better by a constant increment, that Wednesday will be better than Tuesday, and Tuesday will be better than Monday. The curve is really one of an ascending sawtooth: there are setbacks, and there's no magical law that raises us ever higher. But on average the long-term trajectory is unmistakable, even if there are local delays.

RUDYARD GRIFFITHS: So you're saying, in a sense, that we shouldn't have nightmares that we'll have a Mayan-style collapse in our future? That disappearing human civilizations is a thing of the past?

STEVEN PINKER: Anything is possible, and only a fool would predict 500, 1,000, 2,000 years out. But, no, I don't think that that's something that we're going to see in our lifetimes, and I don't think there's any reason to think it's particularly probable.

RUDYARD GRIFFITHS: Why is that? Is it because you again see redundancy with complexity? Does this innovation of ideas and problem solving that you mentioned create a more robust system than those that other civilizations functioned with?

STEVEN PINKER: Yes, I think there are a lot of differences between Western civilization in 2015 and the Mayans, such as the fact that we have a whole edifice of science and technology, a means of accumulating our hard-won knowledge. We have created a massive infrastructure of information, and of finance, and education, and government, and other institutions. So it's a bad analogy: the Mayan civilization was very different from ours.

RUDYARD GRIFFITHS: Right. I have a couple of final questions about perspective. People who are not as keen on the idea of progress will say that your viewpoint depends on your historical snapshot. And that yes, right now in the West, we can point to a trajectory of progress going back through the Industrial Revolution, but if you lived in the Dark Ages and you took the thousand years from the fall of Rome to the Renaissance, that might not look like a period of progression in terms of violence rates, life expectancy, and disease mortality. How do you respond to that? Do you feel that these antecedents of progress are larger than just our own industrial era?

STEVEN PINKER: There's been a big change since the scientific revolution and the Enlightenment. It was the point at which innovation really began, or at least accelerated, and it was when we developed the scientific method, which they did not have at the fall of the Roman Empire. This set us on a trajectory of indefinite technological innovation.

So the real watershed was that two-hundred-year

period when we had the scientific revolution and the Enlightenment. I think analogies to non-technological and pre-technological civilizations really don't apply to the world we're living in now.

RUDYARD GRIFFITHS: Interesting. And just finally, I'd like to talk about climate change, because it's on everybody's mind. How do you factor climate change into your argument? Or do you feel that this is a global threat that we'll adapt to and find a way to muddle through? That it's not as existential as many people passionately believe?

STEVEN PINKER: It would be existential if we didn't do anything about it, but I see no reason to believe that we won't. Economists have noted what the solution is: there has to be carbon pricing, so that people will be incentivized to innovate, conserve, and switch to low-carbon energy sources; and there has to be R&D in transformational technologies, such as carbon capture, battery technology, and a new generation of nuclear power. If you combine those, then dealing with climate change will still be a formidable challenge—it might be the greatest challenge that we've ever faced—but it is not an unsolvable problem, and it's not one that the world is likely to suicidally ignore.

RUDYARD GRIFFITHS: It's not an existential crisis.

STEVEN PINKER: It would be if we don't do anything about it, but there is no reason to think that we won't do anything about it.

RUDYARD GRIFFITHS: Right. The incentives are there to do something—

STEVEN PINKER: Yeah.

RUDYARD GRIFFITHS: In a big way. Steven Pinker, thank you so much for coming to Toronto for this debate. You've been a really important international participant in the conversation, and it's great to have you here.

STEVEN PINKER: Thank you for having me.

Post-Debate Commentary

POST-DEBATE COMMENTARY BY ALI WYNE

At the core of this debate are disagreements about the nature of progress and the soundness of inference. In brief, Steven Pinker and Matt Ridley focus primarily on high-level, favourable post-war trends in peace, health, and prosperity. They believe that on account of rapid, far-reaching innovations in "things," "rules," and "tools" (Ridley's breakdown), there is good reason to believe those trends can and will be sustained. Malcolm Gladwell counters that one cannot evaluate these phenomena one way or another unless one knows what other trends and events they have caused — and will produce: How is one supposed to conduct such a net assessment? Gladwell is also deeply suspicious of any attempts to project the future. Alain de Botton — interestingly, if ambiguously — defines progress as an increase in wisdom. He takes an instrumentally dim view of humanity and its prospects, arguing that pessimism about the future is a vital guarantee of preparedness and anticipation.[1]

Pinker has arguably done more than any of his contemporaries to stimulate a debate about the trajectory of violence—in all its expressions. In his 2011 opus *The Better Angels of Our Nature*, he concludes that "violence has declined over long stretches of time, and today we may be living in the most peaceable era in our species' existence." Pinker concedes that the number of civil wars has increased in recent years—from four in 2007 to eleven in 2014; he also notes that the implosion of Syria has caused the global rate of battle deaths to increase slightly—from less than 0.5 between 2001 and 2011 to 1.4 last year. But these figures are a far cry from previous levels: there were 26 civil wars in 1992, and the global rate of battle deaths hovered around five in the mid-1980s.[2] Interstate wars and mass killings of unarmed civilians, meanwhile, have fallen dramatically over the past seven decades. Leaving aside both methodological disagreements and the question of whether the postwar trends Pinker documents are aberrant, it would seem to require impressive intellectual acrobatics to deny that humanity is, on balance, becoming less violent.

Ridley, a self-described "rational optimist," cites the

---

[1]   Mark Medley, "Munk Debate: Be it resolved humankind's best days lie ahead," *Globe and Mail*, November 6, 2015, . http://www.theglobeandmail.com/news/national/munk-debate-be-it-resolved-humankinds-best-days-lie-ahead/article27134371/.

[2]   Steven Pinker, "Now for the good news: Things really are getting better," *Guardian*, September 11, 2015, http://www.theguardian.com/commentisfree/2015/sep/11/news-isis-syria-headlines-violence-steven-pinker.

human mind's "strange asymmetry — we're very biased in our memories of the past and we're very biased in our assessment of the future." The oft-expressed nostalgia for a halcyon past has little (apparent) basis in fact. After all, interstate wars used to be a reality of life, and the prospect of nuclear Armageddon hung over the world for the second half of the twentieth century (in fact, there were multiple occasions during the Cold War when accidents and misunderstandings nearly brought that nightmarish possibility to pass). Nor is it clear why we would look back fondly upon lower life expectancy and higher rates of illiteracy, poverty, malnutrition, and child mortality. Today's media plays a central role in skewing our perceptions: Ridley observes that "A headline ... tomorrow that world infant mortality went down by .0001 percent yesterday is not, presumably, a good idea in news terms, whereas a headline that an airliner crashed yesterday is much more salient."[3]

As noted earlier, Pinker and Ridley both focus on the big picture. The former notes that "the quantitative data on indicators of human well-being are all pointing in a very consistent, powerful, positive direction — anything from longevity to health to prosperity to peace to democracy." The latter, similarly, cites "the incredible improvements in human living standards over the last fifty years in particular."[4] Gladwell rejects these broad strokes, arguing that one cannot classify a phenomenon as a success

---

[3]   Medley, "Munk Debate."

[4]   Ibid.

unless one can establish that its long-term benefits will outweigh the long-term risks it generates and the damage it wreaks. It is, of course, impossible, to make such an assessment; the best one can do is offer a range of conjectures. For example, while the dramatic reduction in the rate of poverty in China and India over the past forty years is an unquestionably favourable outcome, the industrialization that allowed it to occur has poisoned the countries' air, soil, and water, with consequences that will unfold over generations.

I hasten to note that Gladwell is not a pessimist; in fact, despite his inclusion in this debate, he is agnostic about its central question. One might call him a chastened realist: Gladwell says he has been humbled by the failure of his predictions as well as those of his colleagues. And his caution is warranted: the existence of a trend does not assure its continuation, and progress, however defined, is not guaranteed; it is intrinsically contingent—at least as much upon human intervention as upon unforeseen circumstances. What would be the consequences for world order of an armed confrontation between the United States and China or an act of nuclear terrorism in a major world capital? What new categories of threats will emerge from the advances we are making in biology, robotics, and cyberspace? The United Nations projects that the world's population will grow by 2.4 billion by 2050;[5] given the scale of

---

[5]   Sam Jones and Mark Anderson, "Global population set to hit 9.7 billion people by 2050 despite fall in fertility," *Guardian*, July 29, 2015, http://www.theguardian.com/global-development/2015/jul/29/un-world-population-prospects-the-2015-revision-9-7-billion-2050-fertility.

environmental devastation we have already wrought on our planet, coupled with the pace at which it is accelerating, what effect will this increase have on our forests and oceans, not to mention the arable land whose output feeds vast swathes of the developing world?

Perhaps the most interesting intervention comes from de Botton. He calls the human being "a profoundly flawed creature" who must become "less violent, more forgiving, and more educable" if the world is to have any enduring hope.[6] There is, sadly, too much evidence to recommend his judgements: consider the methods of torture humans have devised, the unknown millions who have perished in their wars, and the waste they have laid to the resources that permit their survival. And it is the misguided conceit of every generation to regard itself as more enlightened and humane than its predecessors. But I am unprepared to reach as damning an indictment as de Botton, for at least two reasons. First, if we abandon all faith in the possibility of redemption — of ourselves and the world we are tasked with protecting — how can we confront its challenges? Second, while humans have caused great destruction, they have also fed untold billions, cured horrific diseases, and redressed enormous injustices. At the risk of speaking prematurely, we should conclude — or at least hope — that the wisdom of our minds and the passion of our hearts scale with the magnitude of our calling.

---

[6]   Medley, "Munk Debate."

Ali Wyne is a nonresident fellow at the Atlantic Council and a security fellow with the Truman National Security Project. He is a co-author of Lee Kuan Yew: The Grand Master's Insights on China, the United States, and the World.

ACKNOWLEDGEMENTS

The Munk Debates are the product of the public-spiritedness of a remarkable group of civic-minded organizations and individuals. First and foremost, these debates would not be possible without the vision and leadership of the Aurea Foundation. Founded in 2006 by Peter and Melanie Munk, the Aurea Foundation supports Canadian individuals and institutions involved in the study and development of public policy. The debates are the foundation's signature initiative, a model for the kind of substantive public policy conversation Canadians can foster globally. Since the creation of the debates in 2008, the foundation has underwritten the entire cost of each semi-annual event. The debates have also benefited from the input and advice of members of the board of the foundation, including Mark Cameron, Andrew Coyne, Devon Cross, Allan Gotlieb, Margaret MacMillan, Anthony Munk, Robert Prichard, and Janice Stein.

For her contribution to the preliminary edit of the book, the debate organizers would like to thank Jane McWhinney.

Since their inception, the Munk Debates have sought to take the discussions that happen at each event to national and international audiences. Here the debates have benefited immeasurably from a partnership with Canada's national newspaper, the *Globe and Mail*, and the counsel of its editor-in-chief, David Walmsley.

With the publication of this superb book, House of Anansi Press is helping the debates reach new audiences in Canada and around the world. The debates' organizers would like to thank Anansi chair Scott Griffin and president and publisher Sarah MacLachlan for their enthusiasm for this book project and insights into how to translate the spoken debate into a powerful written intellectual exchange.

ABOUT THE DEBATERS

STEVEN PINKER is a pioneering cognitive scientist who has written a number of bestselling books, including *The Sense of Style: The Thinking Person's Guide to Writing in the 21st Century*, as well as the landmark study on human progress *The Better Angels of Our Nature*, which won the *New York Times Book Review* Notable Book of the Year Award and was chosen for Mark Zuckerberg's book club. *The Blank Slate* and *How the Mind Works* were both finalists for the Pulitzer Prize. He is Johnstone Family Professor of Psychology at Harvard and has been named by *Time* magazine as one of the "100 Most Influential People in the World."

MATT RIDLEY'S books have been finalists for nine major literary prizes, won several awards, been translated into thirty languages, and sold over one million copies. He currently writes the Mind and Matter column in the *Wall*

Street Journal and contributes regularly to the *Times*. As Viscount Ridley, he was appointed to the House of Lords in 2013 and is a fellow of the Royal Society of Literature, the Academy of Medical Sciences, and a foreign honorary member of the American Academy of Arts and Sciences.

ALAIN DE BOTTON is a writer of essayistic books that have been described as a "philosophy of everyday life." He has written about love, travel, architecture, and literature. His books have been bestsellers in thirty countries. De Botton also started and helps to run a London-based school called The School of Life, dedicated to a new vision of education. His latest book is *The News: A User's Manual*.

MALCOLM GLADWELL is a Canadian journalist and the author of five *New York Times* bestsellers: *The Tipping Point, Blink, What the Dog Saw,* and his latest, *David and Goliath: Underdogs, Misfits, and the Art of Battling Giants.* He has been named one of the "100 Most Influential People in the World" by *Time* magazine and one of *Foreign Policy* magazine's "Top 100 Global Thinkers." Gladwell has been a staff writer for the *New Yorker* since 1996. He has won a National Magazine Award and been honoured by the American Psychological Society and the American Sociological Society.

ABOUT THE EDITOR

RUDYARD GRIFFITHS is the chair of the Munk Debates and president of the Aurea Charitable Foundation. In 2006 he was named one of Canada's "Top 40 under 40" by the *Globe and Mail*. He is the editor of thirteen books on history, politics, and international affairs, including *Who We Are: A Citizen's Manifesto*, which was a *Globe and Mail* Best Book of 2009 and a finalist for the Shaughnessy Cohen Prize for Political Writing. He lives in Toronto with his wife and two children.

ABOUT THE MUNK DEBATES

The Munk Debates are Canada's premier public policy event. Held semi-annually, the debates provide leading thinkers with a global forum to discuss the major public policy issues facing the world and Canada. Each event takes place in Toronto in front of a live audience, and the proceedings are covered by domestic and international media. Participants in recent Munk Debates include Anne Applebaum, Robert Bell, Tony Blair, John Bolton, Ian Bremmer, Stephen F. Cohen, Daniel Cohn-Bendit, Paul Collier, Howard Dean, Hernando de Soto, Alan Dershowitz, Maureen Dowd, Gareth Evans, Mia Farrow, Niall Ferguson, William Frist, Newt Gingrich, David Gratzer, Glenn Greenwald, Michael Hayden, Rick Hillier, Christopher Hitchens, Richard Holbrooke, Josef Joffe, Robert Kagan, Garry Kasparov, Henry Kissinger, Charles Krauthammer, Paul Krugman, Arthur B. Laffer, Lord Nigel Lawson, Stephen Lewis, David Daokui Li,

Bjørn Lomborg, Lord Peter Mandelson, Elizabeth May, George Monbiot, Caitlin Moran, Dambisa Moyo, Vali Nasr, Alexis Ohanian, Camille Paglia, George Papandreou, Samantha Power, Vladimir Pozner, David Rosenberg, Hanna Rosin, Anne-Marie Slaughter, Bret Stephens, Lawrence Summers, Amos Yadlin, and Fareed Zakaria. The Munk Debates are a project of the Aurea Foundation, a charitable organization established in 2006 by philanthropists Peter and Melanie Munk to promote public policy research and discussion. For more information, visit www.munkdebates.com.

ABOUT THE INTERVIEWS

Rudyard Griffith's interviews with Malcolm Gladwell, Alain de Botton, Matt Ridley, and Steven Pinker were recorded on November 6, 2015. The Aurea Foundation is gratefully acknowledged for permission to reprint excerpts from the following:

(p. 63) "Malcolm Gladwell in Conversation," by Rudyard Griffiths. Copyright © 2016 Aurea Foundation. Transcribed by Transcript Divas.

(p. 69) "Alain de Botton in Conversation," by Rudyard Griffiths. Copyright © 2016 Aurea Foundation. Transcribed by Transcript Divas.

(p. 77) "Matt Ridley in Conversation," by Rudyard Griffiths. Copyright © 2016 Aurea Foundation. Transcribed by Transcript Divas.

ABOUT THE POST-DEBATE COMMENTARY

Ali Wyne's post-debate commentary was written following the debates on November 6, 2015. The Aurea Foundation wishes to thank Rudyard Griffiths for his assistance in commissioning this essay.

Should the West Engage Putin's Russia?
Cohen and Pozner vs. Applebaum and Kasparov

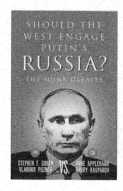

How should the West deal with Vladimir Putin? Acclaimed academic Stephen F. Cohen and veteran journalist and bestselling author Vladimir Pozner square off against internationally renowned expert on Russian history Anne Applebaum and Russian-born political dissident Garry Kasparov to debate the future of the West's relationship with Russia.

"A dictator grows into a monster when he is not confronted at an early stage...And unlike Adolf Hitler, Vladimir Putin has nuclear weapons."
—Garry Kasparov

Has Obama Made the World a More Dangerous Place?

Kagan and Stephens vs. Zakaria and Slaughter

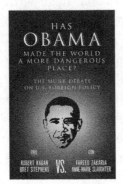

From Ukraine to the Middle East to China, the United States is redefining its role in international affairs. Famed historian and foreign policy commentator Robert Kagan and Pulitzer Prize–winning journalist Bret Stephens take on CNN's Fareed Zakaria and noted academic and political commentator Anne-Marie Slaughter to debate the foreign policy legacy of President Obama.

"Superpowers don't get to retire ... In the international sphere, Americans have had to act as judge, jury, police, and, in the case of military action, executioner." —Robert Kagan

Does State Spying Make Us Safer?
Hayden and Dershowitz vs. Greenwald and Ohanian

In a risk-filled world, democracies are increasingly turning to large-scale state surveillance, at home and abroad, to fight complex and unconventional threats. Former head of the CIA and NSA Michael Hayden and civil liberties lawyer Alan Dershowitz square off against journalist Glenn Greenwald and reddit co-founder Alexis Ohanian to debate if the government should be able to monitor our activities in order to keep us safe.

"Surveillance equals power. The more you know about someone, the more you can control and manipulate them in all sorts of ways."
—Glenn Greenwald

Are Men Obsolete?
Rosin and Dowd vs. Moran and Paglia

For the first time in history, will it be better to be a woman than a man in the upcoming century? Renowned author and editor Hanna Rosin and Pulitzer Prize–winning columnist Maureen Dowd challenge *New York Times*–bestselling author Caitlin Moran and trailblazing social critic Camille Paglia to debate the relative decline of the power and status of men in the workplace, the family, and society at large.

"Feminism was always wrong to pretend women could 'have it all.' It is not male society but Mother Nature who lays the heaviest burden on women." —Camille Paglia

Should We Tax the Rich More?
Krugman and Papandreou vs. Gingrich and Laffer

Is imposing higher taxes on the wealthy the best way for countries to reinvest in their social safety nets, education, and infrastructure while protecting the middle class? Or does raising taxes on society's wealth creators lead to capital flight, falling government revenues, and less money for the poor? Nobel Prize–winning economist Paul Krugman and former prime minister of Greece George Papandreou square off against former speaker of the U.S. House of Representatives Newt Gingrich and famed economist Arthur Laffer to debate this key issue.

"The effort to finance Big Government through higher taxes is a direct assault on civil society."   —Newt Gingrich

Can the World Tolerate an Iran with Nuclear Weapons?
Krauthammer and Yadlin vs. Zakaria and Nasr

Is the case for a pre-emptive strike on Iran ironclad? Or can a nuclear Iran be a stabilizing force in the Middle East? Former Israel Defense Forces head of military intelligence Amos Yadlin, Pulitzer Prize–winning political commentator Charles Krauthammer, CNN host Fareed Zakaria, and Iranian-born academic Vali Nasr debate the consequences of a nuclear-armed Iran.

"Deterring Iran is fundamentally different from deterring the Soviet Union. You could rely on the latter but not the former."
—Charles Krauthammer

Has the European Experiment Failed?
Joffe and Ferguson vs. Mandelson and Cohn-Bendit

Is one of human history's most ambitious endeavours nearing collapse? Former EU commissioner for trade Peter Mandelson and EU Parliament co-president of the Greens/European Free Alliance Group Daniel Cohn-Bendit debate German publisher-editor and author Josef Joffe and renowned economic historian Niall Ferguson on the future of the European Union.

"For more than ten years, it has been the case that Europe has conducted an experiment in the impossible."

—Niall Ferguson

North America's Lost Decade?
Krugman and Rosenberg vs. Summers and Bremmer

 The future of the North American economy is more uncertain than ever. In this edition of the Munk Debates, Nobel Prize–winning economist Paul Krugman and chief economist and strategist at Gluskin Sheff + Associates David Rosenberg square off against former U.S. treasury secretary Lawrence Summers and bestselling author Ian Bremmer to tackle the resolution, "Be it resolved: North America faces a Japan-style era of high unemployment and slow growth."

"It's now impossible to deny the obvious, which is that we are not now, and have never been, on the road to recovery."
— Paul Krugman

Does the 21st Century Belong to China?
Kissinger and Zakaria vs. Ferguson and Li

Is China's rise unstoppable? Former U.S. secretary of state Henry Kissinger and CNN's Fareed Zakaria pair off against leading historian Niall Ferguson and world-renowned Chinese economist David Daokui Li to debate China's emergence as a global force — the key geopolitical issue of our time.

This edition of the Munk Debates also features the first formal public debate Dr. Kissinger has participated in on China's future.

"I have enormous difficulty imagining a world dominated by China... I believe the concept that any one country will dominate the world is, in itself, a misunderstanding of the world in which we live now." —Henry Kissinger

Hitchens vs. Blair
Christopher Hitchens vs. Tony Blair

Intellectual juggernaut and staunch atheist Christopher Hitchens goes head-to-head with former British prime minister Tony Blair, one of the Western world's most openly devout political leaders, on the age-old question: Is religion a force for good in the world? Few world leaders have had a greater hand in shaping current events than Blair; few writers have been more outspoken and polarizing than Hitchens.

Sharp, provocative, and thoroughly engrossing, *Hitchens vs. Blair* is a rigorous and electrifying intellectual sparring match on the contentious questions that continue to dog the topic of religion in our globalized world.

"If religious instruction were not allowed until the child had attained the age of reason, we would be living in a very different world." —Christopher Hitchens

houseofanansi.com/collections/munk-debates

The Munk Debates: Volume One
Edited by Rudyard Griffiths; Introduction by Peter Munk

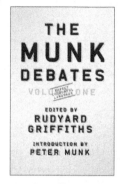

Launched in 2008 by philanthropists Peter and Melanie Munk, the Munk Debates is Canada's premier international debate series, a highly anticipated cultural event that brings together the world's brightest minds.

This volume includes the first five debates in the series and features twenty leading thinkers and doers arguing for or against provocative resolutions that address pressing public policy concerns, such as the future of global security, the implications of humanitarian intervention, the effectiveness of foreign aid, the threat of climate change, and the state of health care in Canada and the United States.

"By trying to highlight the most important issues at crucial moments in the global conversation, these debates not only profile the ideas and solutions of some of our brightest thinkers and doers, but crystallize public passion and knowledge, helping to tackle some global challenges confronting humankind."

—Peter Munk

houseofanansi.com/collections/munk-debates

CPSIA information can be obtained
at www.ICGtesting.com
Printed in the USA
LVOW03s1249220218
567346LV00004B/4/P